This book defends the prospects for a science of society. It argues that behind the diverse methods of the natural sciences lies a common core of scientific rationality that the social sciences can and sometimes do achieve. It also argues that good social science must be in part about large-scale social structures and processes and thus that methodological individualism is misguided. These theses are supported by a detailed discussion of actual social research, including theories of agrarian revolution, organizational ecology, social theories of depression, and supply-and-demand explanations in economics.

Professor Kincaid provides a general picture of explanation and confirmation in the social sciences and discusses the nature of scientific rationality, functional explanation, optimality arguments, meaning, and interpretation; the place of microfoundations in social explanation; the status of neo-classical economics; the role of idealizations and non-experimental evidence; and other controversies in social research.

PHILOSOPHICAL FOUNDATIONS OF THE SOCIAL SCIENCES

Philosophical foundations of the social sciences

ANALYZING CONTROVERSIES IN SOCIAL RESEARCH

HAROLD KINCAID
University of Alabama at Birmingham

CAMBRIDGE UNIVERSITY PRESS

Published by the Press Syndicate of the University of Cambridge
The Pitt Building, Trumpington Street, Cambridge CB2 1RP
40 West 20th Street, New York, NY 10011–4211, USA
10 Stamford Road, Oakleigh, Melbourne 3166, Australia

© Cambridge University Press 1996

First published 1996

Library of Congress Cataloging-in-Publication Data
Kincaid, Harold, 1952–
    Philosophical foundations of the social sciences : analyzing controversies in social research / Harold Kincaid.
        p. cm.
    ISBN 0-521-48268-2. — ISBN 0-521-55891-3 (pbk.)
    1. Social sciences—Research. 2. Social sciences—Philosophy.
  I. Title.
H62.K515 1996
300′.1—dc20   95–13774
                                                            CIP

A catalog record for this book is available from the British Library

ISBN   0-521-48268-2 Hardback
        0-521-55891-3 Paperback

Transferred to digital printing 2001

In memory of the generation that went before

Harold Wilson Kincaid
1920–1943

Earl Alexander Kincaid, Jr.
1926–1970

Katherine Kincaid Brooks
1919–1987

CONTENTS

List of figures	*page* xi
Acknowledgments	xiii
Preface	xv
Chapter 1: Issues and arguments	1
1.1 The naturalist and holist traditions and their detractors	2
1.2 An outline of the argument	9
Chapter 2: Challenges to scientific rationality	16
2.1 Quine and the demise of positivism	17
2.2 Varieties of rationality	27
2.3 Kuhn and shifting standards	30
2.3.1 Incommensurability	30
2.3.2 Theory-laden data	33
2.3.3 Ambiguous criteria	35
2.4 Social constructivism and post-modernist rhetoric	37
2.5 The subtle invasion of values	43
2.6 The symptoms of good science	47
Chapter 3: Causes, confirmation, and explanation	58
3.1 Some *a priori* objections	59

viii   Contents

3.2 Confirmation and qualifications	63
3.2.1 How can *ceteris paribus* laws be confirmed and how can they explain?	63
3.2.2 *Ceteris paribus* in practice	70
3.3 Inferring causes from non-experimental data	84
3.4 Lawless explanations	90
Chapter 4: Functionalism defended	**101**
4.1 Functionalism and its critics	103
4.2 What is functionalism?	105
4.3 Confirming functional explanations	114
4.4 Functionalist failures and successes	122
4.4.1 Optimal Eskimos and Hindu Cows	122
4.4.2 Marxist accounts of the state	126
4.4.3 The ecology of organizations	131
4.5 The critics answered	136
Chapter 5: The failures of individualism	**142**
5.1 The prospects for reduction	145
5.1.1 Requirements for reduction	145
5.1.2 Conceptual arguments for and against reducibility	149
5.1.3 An empirical case against individualism	153
5.2 Claims about explanation and confirmation	166
5.2.1 Full explanation without reduction?	167
5.2.2 Is individualism the *best* explanation?	175
5.2.3 Are individualist *mechanisms* necessary?	179
5.2.4 Individualist evidence and heuristics	182
5.3 A question of ontology?	187
5.4 The truth in individualism	189
Chapter 6: A science of interpretation?	**191**
6.1 Issues and presuppositions	192

Contents   ix

6.2 The right-wing attack   194
6.3 Skeptical hermeneuts   205
6.4 Interpretive successes   212
6.5 Norms and symbols   215

Chapter 7: Economics: a test case   222
7.1 How to think about economics   223
7.2 The supply-and-demand core   232
    7.2.1 Confirming the laws of supply and demand   232
    7.2.2 The central role of supply-and-demand
        arguments   241
7.3 Assessing neo-classical models   247
7.4 Reduction and microfoundations   250

Chapter 8: Problems and prospects   258

References   266
Index   279

FIGURES

1: Paige's four basic agrarian class systems.	*page* 71
2: The political behavior of non-cultivators.	72
3: The political behavior of cultivators.	74
4: Agrarian classes and political behavior.	75
5: Paige's correlations.	87
6: Homeostatic model of functional explanations.	107
7: Complex functional models.	121
8: Examples of spurious correlations.	129
9: Hannan and Freeman's hypotheses.	133

ACKNOWLEDGMENTS

This book took shape over many years, and I have incurred numerous debts in the process. Philosophers and social scientists who take time to respond carefully to their colleagues' work in the rough do a great service, and I am enormously grateful for the help that I have had. Alexander Rosenberg, Wade Hands, Merilee Salmon, John Dupre, Daniel Little, Paul Churchill, and several anonymous referees read the entire manuscript and made numerous valuable comments. Barbara Horan, Andrew Vayda, Alan Nelson, Dan Hausman, and Mark LaGory provided detailed comments on specific chapters. My colleagues Timothy Day and Scott Arnold likewise commented on specific chapters; more important, they were a ready source of both philosophical advice and friendship during the trials and tribulations of book writing. Robert Causey, Paul Teller, John Post, Terry Horgan, David Henderson, Lee McIntyre, Bob McCauley, George Graham, Graham McDonald, Marthe Chandler, and Neil de Marchi made many useful suggestions about the basic arguments of the book as they first appeared in various journal articles. Peter Urbach did so as well; he was also a delightful host and friend during my visit to LSE while I was writing this book.

I also owe a great debt to two teachers and friends, Scott Gordon and Geoffrey Hellman. Scott provided substantive comments on Chapter 7; Geoffrey provided extensive feedback on my earliest papers on individualism. My debt to them, however, goes far deeper. What rigor I am able to muster in this book is largely due to Geoffrey's influence; what I know of economics and how I approach the philosophy of social science I largely owe to Scott. No doubt my work is all too often a pale reflection of the original, but a reflection it is.

At Cambridge University Press, Scott Parris has been an efficient and intelligent editor; W. M. Havighurst saved me from numerous errors, both large and small, and the manuscript is much better because of his careful copy editing. I thank them both for their efforts.

I also thank Sage Publications for permission to reprint parts of my "Defending Laws in the Social Sciences," *Philosophy of the Social Sciences* 20 (1990): 56–83, and the Philosophy of Science Association for permission to reprint parts of the following: "Confirmation, Complexity, and Social Laws," *PSA 1988,* vol. 2, ed. A. Fine and J. Leplin, pp. 299–307; "Explanation, Reduction, and Individualism," *Philosophy of Science* 53 (1986): 343–356; and "Assessing Functional Explanation in the Social Sciences," *PSA 1990,* vol. 1, ed. A. Fine, M. Forbes, and L. Wessels, pp. 341–354.

Finally, thanks to Henry and Gaylia Levkoff for producing the figures for the book and for their unfailing willingness to eat Indian food in a construction site. And to Robin (and Murphy) I owe more than I can say.

PREFACE

This book results from three convictions: (1) that pressing social problems such as poverty, discrimination, and inequality are not simply the result of individual characteristics but result instead from larger social structures; (2) that scientific methods are the most powerful tools available for replacing superstition and prejudice with knowledge and thus that we can and ought to study those social structures with the methods of the natural sciences, broadly construed; and (3) that philosophy of science can contribute to developing such a science of society, but only if it eschews *a priori* armchair theorizing in favor of a philosophy intimately tied to the real practice of social science research. The first two convictions have their origin in the Enlightenment. However, the more direct cause in my case was growing up in an era flush from the discoveries of Watson and Crick and a personal involvement in the political movements of the 1960s and 1970s. The third, more cerebral conviction has more recent origins. It results not only from the philosophical arguments of Kuhn and Quine but also from a growing frustration with my philosophical colleagues who are willing to pronounce entire domains of social inquiry doomed to failure while paying little attention to what social scientists actually do.

In one way or another every discussion in this book flows from and serves to reinforce these three convictions. My faith that scientific methods can give us knowledge of large scale social processes produces the two big theses I defend in this book, namely, naturalism and holism. Behind the diverse methods of the natural sciences lies, I believe, a common core of scientific rationality that the social sciences can and sometimes do achieve. Naturalism is thus the belief that social phenomena are part of the natural world and accordingly amenable to the methods of the natural sciences. Holism is the doctrine that good social science will be at least in part about the large scale and that methodological individualism is thus misguided. One goal of this book is to defend these grand theses about the social sciences.

That defense, however, will not come from some deep philosophical insight into the nature of human beings, the mental, or whatever. The "philosophical foundations" I address are contingent and empirical, not truths from outside of scientific practice. Thus, I argue for the naturalist and holist positions by looking in detail at current research and controversies in the social sciences. Determining the prospects for good social science requires looking carefully at the kind of theories, explanations, and evidence social scientists produce and then asking in the concrete whether these accounts meet basic standards of scientific adequacy. Similarly, adjudicating the individualism–holism dispute requires sorting out what is ultimately an empirical issue and deciding, insofar as is possible, what the empirical evidence shows. In short, defending the naturalist and holist positions requires analyzing, clarifying, and taking positions on numerous specific controversies in social science research.

The general approach sketched above makes this a book with a specific kind of audience and a particular set of objectives that I should explicitly acknowledge. Since my ultimate hope is to contribute to better social research, this book is written both for social scientists and for philosophers. For social scientists, I hope to show that philosophy of science and philosophical analysis can help clarify important social science controversies – for example, the adequacy of and evidence for functional explanations, the problems raised by *ceteris paribus* assumptions and idealizations, or the strengths and weaknesses of optimal foraging accounts of Inuit hunting practices. In such discussions I have assumed no particular acquaintances with recent work in philosophy of science and consequently have provided background discussions where necessary. I have also aimed for completeness, trying to provide an up-to-date discussion, accessible to social scientists, of all the main topics in the philosophy of the social sciences. For philosophers, I provide an exhaustive criticism of relatively conceptual arguments about what the social sciences can do; a careful sorting out of the issues in standard debates in the philosophy of the social sciences; an embryonic picture of explanation as it works in the social sciences; some new criticisms of Quine's indeterminacy thesis and of unificationist accounts of explanation; and so on. More important, perhaps, I try to help push philosophy of social science towards a much more intimate tie with ongoing social research, a step already taken with great success by our colleagues in philosophy of biology and physics.

1

Issues and arguments

After finishing the first volume of *Capital,* Karl Marx sent a copy of the book along with an admiring note to Charles Darwin. Apparently Darwin found little of interest in Marx's work, for the pages in his copy of *Capital* went uncut. Though Darwin was obviously open to revolutionary ideas in science, he was politically and socially moderate, adhering to the Victorian mores of his upper-class upbringing. Darwin had no desire to be associated with a radical like Marx. Nonetheless, Darwin and Marx did have something very deep in common: a belief that the human species is part of the natural order and thus amenable to scientific understanding. Darwin revolutionized biology and our self-understanding by identifying the processes governing the evolution of species. Marx thought he had done the same for human society. Human society, like any other natural object, was subject to scientific investigation; Marx's task was to "lay bare its laws of motion." He, like Darwin, thought that we could understand the human species, in all its forms, with scientific rigor.

This book defends Marx's grand idea – not his specific thesis of historical materialism but his faith that standard scientific methods can produce scientific knowledge about the large-scale features of society. I thus intend to defend the doctrines sometimes labelled "naturalism" and "holism." Naturalists believe roughly that human society, as part of the natural world, can be understood by the general methods of the natural sciences. Holists think that an adequate social science cannot proceed entirely at the individual level, for macrosociological explanations have an irreducible part to play. In short, I examine the prospects and problems for a science of society.

Although this book focuses on these two large issues, they far from exhaust it. Defending naturalism and holism requires examining and engaging in numerous more specific controversies in current social research.

Philosophy of science has learned in the last thirty years that it cannot proceed apart from actual scientific practice and controversy. Broad philosophic theses about science are in the end part of science itself. Thus *a priori* conceptual arguments will not suffice to evaluate naturalism and holism in the social sciences; only careful attention to the ins and outs of social scientific practice will do. Philosophy of physics and biology, for example, is now tightly connected to ongoing empirical work.[1] Philosophers of the social sciences, however, have been slower to get the message.[2] One goal of this book is to help change that situation.

Although I cannot entirely avoid conceptual arguments about what the social sciences could never do or must be able to do (since philosophers keep advancing them), my main claims depend crucially on analyzing empirical work and controversies in the social sciences. I want to show that a social science is not only possible but to some extent actual – in other words, that the obstacles to a science of society are ordinary practical obstacles that can be and sometimes are overcome. As a result, pursuing my main theses will likewise involve analyzing and engaging in concrete empirical debates in social research. So this book is as much about naturalism and holism as it is about debates over symbolist anthropology, theories of the firm, the role of microfoundations in macroeconomics, the validity of survey research, the theory of consumer demand, optimality arguments in the social sciences, organizational ecology, inferring causes from aggregate evidence, rational choice theory, and so on. I regard these specific analyses as important as the "big" theses of naturalism and holism.

This first chapter surveys the basic issues and arguments of the book. I explain why the debate over naturalism and holism matters and what motivates these doctrines. I also survey the general threats that confront any argument for a science of society. I also outline the main arguments chapter by chapter. In the process the empirical controversies that surface throughout the book are previewed. Later chapters work hard to sort out issues; here I focus on the broad sweep of the argument.

1.1 The naturalist and holist traditions and their detractors

The debate over naturalism did not begin nor end with Marx. Marx's commitment to a science of society simply extended an Enlightenment tenet. Just as reason could penetrate a complex natural world, so

---

[1] I have in mind work such as Sober (1984) and Kitcher (1985) in philosophy of biology, Michael Friedman (1983) in physics, and many other recent contributions in these areas.

[2] There are of course some notable exceptions, not limited to but including: Rosenberg (1976), Martin (1989), Little (1989), and Hausman (1981b, 1992).

too could it analyze and evaluate human affairs. Human behavior and culture, carefully investigated, reveal patterns and laws. What Newton did for the natural world, Montesquieu, Stewart, and Smith thought could be done for the social world. Durkheim, Weber, Radcliffe-Brown, and others carried on the tradition after Marx.

What exactly does this naturalist tradition hold about the social sciences? Later chapters will flesh out naturalism in detail, but it will be helpful to have an initial statement here. The naturalism I defend asserts *roughly* that:

(1) the social sciences can be good science by the standards of the natural sciences;
(2) the social sciences can only be good science by meeting the standards of the natural sciences.

My main concern shall be with the first thesis. It argues that the canons of scientific rationality can guide our study of the social realm. Of course, actual social research does and will continue to employ methods found nowhere in the natural sciences and vice versa. But in concrete research the different natural sciences employ diverse methods as well. Behind those different practices lies a commitment to certain basic scientific virtues. Naturalists believe those virtues can also undergird the methods of social research.

The idea that the social sciences *can* be good science has several interpretations, depending on how we understand "can." Many philosophers and social scientists think that no social science is possible; there are, in their view, conceptual considerations ruling out naturalism. Others allow that science in the social sciences is possible in principle but doubt that it is in practice. My aim is to argue with both. Social science can in principle and in practice achieve the basic virtues of the natural sciences. I shall argue for this claim quite directly. No conceptual arguments show social science in principle impossible; some social research shows that the social sciences sometimes achieve full scientific rigor.

Thus the naturalism I advocate holds that some social science is good science. Note, however, that naturalism does not imply that all is well in social research. It would be easy enough to defend naturalism simply by defining good science very broadly – as, for example, "a concern for the facts," something the worst social science shares with our best physics. Yet that defense would both be trivial and make us mere cheerleaders for whatever social scientists happen to produce. The naturalism presented here has no such implications. Large parts of the social sciences do fail to produce good science. Still, that situation can be and sometimes is overcome.

The second naturalist thesis I shall defend asserts that good social science must follow the standards of the natural sciences. This claim is apparently the more controversial of the two, since it tells us not just what can happen but what must. However, how we understand this "must" will determine how radical this claim really is. Legislating what science must be like is risky business – and a business with a long history of failure. One important theme of this book is that we cannot decide the fate of the social sciences on *a priori* conceptual grounds. Thus I shall not argue that good social sciences could never result from methods entirely unknown to the natural sciences. We cannot rule out the possibility that the social sciences might teach us something about fundamental scientific virtues. Indeed, the natural sciences have often learned specific investigative techniques from the social sciences.[3]

How then can we say that the social sciences *must* proceed by the standards of the natural sciences? We can only make a guarded claim based on what we know about social scientific practice as it exists now and its likely permutations. Nothing about current social research shows that good social science proceeds by standards and methods not found in the natural sciences. Social scientists sometimes claim that social phenomena call for entirely unique routes to confirmation and explanation. Their claims, I shall argue, are unfounded. Neither the meaningful nature of social phenomena nor their explanation by appeal to functions nor anything else shows that current social science has its own, entirely special route to knowledge. So based on what we know about the social sciences now, we have no reason to think that the social sciences can proceed without meeting the basic standards of the natural sciences. However unique the methods of good social science are, they still embody the basic virtues driving the natural sciences.

This debate over naturalism is not just a scholastic quarrel among philosophers. There are deep and vitally important questions at stake. If naturalistic methods cannot explain society, then the scientific world view is radically incomplete, a surprising conclusion indeed. Moreover, the Enlightenment thinkers who popularized naturalism did so for more than purely intellectual reasons. They thought that knowledge was power – that understanding how society worked would tell us how to change it for the

---

[3] See, for example, the use of social science statistical methods in biological debates over group selection (Damuth and Heiser 1988) or more generally the application of game theory and models of constrained optimization from economics to evolutionary biology.

better. Thus the question whether the social sciences can be real sciences is far from sterile, for its outcome has enormous consequences, both theoretical and practical.

Consider first the intellectual consequences. Ignoring for the moment currently trendy forms of skepticism, science constitutes our best chance at limning the true structure of reality. If any knowledge tells us the way the world is, it is scientific knowledge. Thus the label "scientific" is not just honorific. If the social sciences can be significantly like our best science, then they tell us something important about reality. If, on the other hand, the social sciences really more closely approximate literature and the humanities, then they take on a different status. They may serve useful functions, but describing the basic patterns of nature will not be one of them.

Furthermore, if a science of human social behavior is impossible, that leaves us with an intellectual puzzle and an existential problem. The puzzle is why our most successful intellectual enterprise should be of no use when it comes to what we care most about, namely, ourselves. The existential problem is that if naturalism is false, we apparently have little hope of understanding ourselves. Human life is permeated with social processes. If we cannot find reasonable evidence about the causes of social events, we have little chance of fully understanding who we are, where we have been, or where we are going.

Our attitude towards a naturalistic social science also has serious implications for what social scientists should be doing. If social science is impossible, then the search for well-confirmed causal generalizations and the like are a waste of time. The debate over naturalism is thus also a debate over how social research should proceed.

Finally, deciding the status of the social sciences has consequences for the other sciences. If no real social science is possible, then biology need not cohere with the claims of sociology. In fact, biology in the form of sociobiology would probably become our best chance for understanding human social behavior. However, if the social sciences provide real scientific understanding, then things are much different. Biological accounts of human behavior must be consistent with – and, better yet, integrated with – those of the social sciences, just as biology must be integrated with chemistry. And sociobiology would not win the day by default. So naturalism says much about how we should try to understand the world.

The practical consequences run equally deep. If no social science is possible, if the best that social scientists can do is give us many different kinds of literary "thick" description of social reality, then social policy is groundless. Imagine that we could have no real knowledge about social processes. Government intervention in social and economic affairs would be inane. How could we evaluate educational programs, prison reform, economic policy and so on without having well-confirmed generalizations about the causes of the social phenomena? We could not. Without knowledge of causes, we have no idea what factors should be manipulated or in which direction. Policy making would be guessing in the dark. Likewise, social movements to eliminate racism and sexism would be misguided, for they would have no rational basis for acting. Although many critics of naturalistic social science come from the left, rejecting naturalism really has quite conservative implications. Of course, even a real social science might not give us a basis for action: we might learn that some social problems are practically ineliminable. But, obviously, even that information would also make important differences to policy.

Defending science in the social sciences is only part of my project. At some level of description we surely already have a science of human behavior – physics and biology after all generate numerous predictions about humans and their interactions (we cannot communicate faster than the speed of light and every society depends on some form of agriculture!). Of course those predictions are not very interesting qua social science. For the classical tradition of Marx and Durkheim, much the same holds for the predictions of psychology: they may tell us something about human behavior and interaction, but they do not constitute a science of *society*. A real social science would provide us with laws, predictions, and understanding of large-scale social structure. Social science should describe how institutions relate to and influence one another, how social structures develop and change, and how those institutions and structures influence the fate of individuals. In short, social science is in some sense autonomous from psychology and has its own domain of inquiry.

Like naturalism, this holist tradition makes claims about how the social sciences *can* and *must* proceed. In the version I defend, holism claims *roughly* that:

(1) there can be good social science that explains in terms of social entities – such as classes, institutions, and so on – and their characteristics;
(2) the social sciences must explain partly in terms of social entities and their characteristics.

The first thesis denies that there is any inherent or conceptual flaw in a holist social science. Explanations that invoke macrosociological entities and processes can be perfectly respectable science. As with the naturalist theses, I take the "can" as a claim about what is possible both in principle and in practice. There are, I shall argue, neither conceptual nor insurmountable practical obstacles to successful holist social science. Establishing that claim requires both answering *a priori* arguments from the critics and arguing that some holist social research succeeds.

The second thesis claims that holist social theory has an essential place. I will not claim that *all* social science must invoke macrosociological processes. However, I will argue that large and important parts of the social sciences must proceed at least in part holistically. The "must" here is again not an eternal conceptual truth about all future social science. Rather, it is an empirical claim based on what we know about the social sciences so far and what we can reasonably expect from them given that experience.

The holist view of social science which I defend thus denies the quite popular doctrine of methodological individualism. Individualism is a fuzzy doctrine. Sometimes it makes ontological claims, for example, that social entities do not exist or act independently of their parts. Other individualists put the issues in terms of knowledge: we can capture all social explanations in individualist terms or no social explanation is complete or confirmed without individualist mechanisms. These holist claims of course also need to be clarified, for they still hide important ambiguities. I shall begin distinguishing various holist and individualist claims below and do so in great detail in Chapter 5. At this point, however, let me explain why the issue matters.

The debate over holism and individualism, I shall argue, is primarily an empirical issue over how to explain society. Holists deny that purely individualist accounts of social structure are ever fully adequate. If the holists are right, then important conclusions follow. It would be wrong to demand that social science ought to proceed individualistically, for "ought" implies "can." A holist like Durkheim, for example, was not simply a mystic believer in some collective spirit. Rather, he thought there was good evidence that individual motives varied too greatly to account for large-scale social patterns (Durkheim 1965). So if there was to be a social science at all, it had to be holistic. Modern-day holists argue along with Durkheim that individualist theories are empirically inadequate. Yet, much contemporary social science makes methodological individualism its official methodology. Thus if holism is correct, then many social scientists follow a misguided philosophy or do not practice what they preach. The holism–individualism debate goes to the core of how social science should proceed.

Important normative and philosophical issues are also at stake. Individualist explanations lend credence to voluntaristic, merit-based political philosophies; holist accounts suggest a very different picture. In addition, we must decide the holism-individualism issue if we want to know how the sciences cohere. If individualism is true, we have support for the positivist idea that there is really only one big science, namely, physics. Holism argues for a more complex – or, from the holist's viewpoint, more sophisticated – account of how the sciences fit together.

Doubts about naturalism and holism come from several directions. The idea that social science can be real science is attacked from, as it were, both the right and the left – by those who want to defend traditional scientific standards and by those who see them as irrelevant or even pernicious. Much of this book thus deals with those threats.

From the left the complaint is that naturalistic social science is a form of scientism or science worship. The interpretivist or hermeneutical tradition denies that we can understand human behavior in terms of laws, causes, and predictions – and likewise denies that this is regrettable. Human behavior is meaningful, and that makes a traditional *science* of society impossible (Geertz 1973b; Taylor 1980; Dilthey 1989). We can understand social phenomena, but not by natural science methods. The human sciences need to grasp the meaning of behavior, and they have their own methods for such an interpretive enterprise. Those methods are the ones appropriate to a hermeneutical activity, not a naturalistic science. Many practicing social scientists see their discipline in just this light. Similar criticisms come from those who deny that the social sciences can be value neutral. Describing human society, so the argument goes, essentially presupposes some idea of what it ought to be like. So an objective, descriptive account of society is a pipe dream.

An even more radical "left-wing" criticism comes from currently trendy forms of irrationalism. Advocates of the social studies of knowledge (Bloor 1976) or of science as rhetoric (McCloskey 1985) as well as neopragmatists like Rorty (1982a), to name a few, all deny that there is something special about science at all. Science is just one kind of conversation or one form of social organization. Science has no special method, no better chance of finding the truth, no privileged form of justification. If the natural sciences have no special claim to rationality, then we need not worry whether the social sciences can be like them. The irrationalists' claims would therefore make the whole issue of naturalism moot. Even those who do not follow Kuhn and company to their irrationalist conclusions may nonetheless think there are important truths here. After all, haven't we learned that the positivist attempt to demarcate science from pseudoscience was a failure? And thus that it is confused to ask whether some discipline uses the scientific method? Obviously, the irrationalists have to be answered as well as the interpretivists.

The right-wing attack on social science comes from those who, to paraphrase Glymour (1980, p. ix), would rather be logical positivists than goddamned English professors. The logical positivists may have had an overly simple conception of science, but they were right that verification, falsifiability, and other such traits are symptoms of good science – and that there is an important difference between science and pseudoscience. However, all indications are that contemporary social science does not and cannot have those traits. Social science, they claim, is a misnomer. Obviously this challenge must be answered.

If no social science is possible, then clearly no holistic social science is either. However, many believe that a holistic social science is problematic even if good social science is possible. Methodological individualists have advanced numerous ontological and conceptual reasons for this conclusion. The more interesting challenges to holism, however, are empirical. For example, economists have been busy applying game theory and rational choice theory to non-economic social phenomena (Becker 1976a). If successful, the economic approach threatens to do in practice what individualists always claimed must be doable in theory: completely explain social phenomena in terms of individuals and their beliefs. To defend holism, I must address these individualist programs as well as the more general conceptual arguments.

Though holism and naturalism are independent theses, they naturally reinforce each other. Defending holism indirectly supports naturalism. Much social science proceeds at the macrolevel. So doubts about holism fuel doubts about naturalism. In turn, defending the prospects for naturalism makes holism more reasonable as well. Most macrosociological research employs naturalist methods. Removing doubts about that approach is thus also a first step in defending holism.

1.2 **An outline of the argument**

The chapters that follow have a common structure. They first explicate and then criticize the relatively *a priori* arguments alleging that no science of society is possible. After removing conceptual obstacles, I defend a more interesting and controversial thesis: that specific pieces of social research meet basic standards of scientific adequacy and/or support the holist conception. The moral is that the only obstacles to a science of society are practical and eliminable ones. It is in arguing for this stronger position that I take up the many specific controversies in social research mentioned earlier; these latter discussions should stand on their own as clarifications of social scientific practice.

Chapter 2 discusses general issues in the philosophy of science. It does so in part for those social science readers who are inclined to think that Popper and Kuhn are the latest word in philosophy of science. They are not, and some common errors result because critics of naturalism think otherwise. Those who give general conceptual arguments denying a science of society frequently appeal to an outdated and inadequate understanding of how natural science operates. Thus this chapter will provide important background for the ones that follow. Perhaps more important, Chapter 2 sketches an answer to those who deny the rationality of science. I try to show how we can reject the relativist conclusions of Kuhn, Rorty, and the social constructivists. My main line of attack – which draws on the work of many others, beginning fundamentally with Quine – is that the irrationalists have an overly simplified picture of science. Attention to the real detail of scientific practice shows that translation across theories is possible, that scientific standards can be rationally criticized, that theory does not necessarily permeate data in a vicious way, and that the world, not just conversation, constrains scientific theory.

These lessons help answer relativist doubts. They also help answer or at least clarify debates about a science of society. In particular, answering the relativists will help us see how there can be basic virtues symptomatic of good science while specific sciences employ quite diverse methods. I divide those basic scientific virtues primarily into those promoting confirmation and those promoting explanation and try to sketch them in enough detail to be informative. Later chapters will rely heavily on those characteristics of good science to argue for naturalism.

Chapter 3 begins the defense proper of science in the social sciences. I focus primarily on the prospect for well-confirmed causal explanations. Many have argued that the social sciences could never produce laws or successful causal explanations. After rejecting those arguments, I move on to a much more interesting and important obstacle. Most causal explanations and generalizations in the social sciences are qualified *ceteris paribus* (other things being equal). Yet in reality other things are seldom equal. We often do not even know what all the potentially confounding factors are, raising the serious worry that social theories are non-falsifiable and non-explanatory. I outline various methods for handling *ceteris paribus* qualifications. Using Paige's (1976) work on the dynamics of agrarian revolutions and Grant's (1986) work in evolutionary biology, I argue that at least some social science deals with the *ceteris paribus* problem as well as some good natural science.

This provides a prima facie case that the social sciences can and sometimes do produce relatively well-confirmed causal explanations. I strengthen that case by looking at doubts about social research based on its non-experimental evidence, on the restricted nature of its generalizations, and on its frequent lack of formalized theories with great unifying power. Non-experimental evidence can be compelling under the right conditions, and those conditions are sometimes met in the social sciences. Though some social research does produce explicit theories that unify diverse phenomena, I argue that it need not to be good science. Much natural science explains without any very elaborate or unifying theories; it is the filling in of causal detail case by case that often does the explanatory work. This is true at least for large parts of biology, and if we take it rather than Newtonian physics as the standard, some social research fares quite well.

Chapter 4 continues the debate over a social science. If any one kind of theory has dominated modern social science, it is functionalism. Functionalism is roughly the doctrine that social institutions exist because of their beneficial effects. Durkheim, Marx, Radcliffe-Brown, and Parsons all espoused some variant of functionalism. Currently, ecological anthropology, human ecology, Marxian sociology, and many other approaches make use of functionalism in some form. Nonetheless, functionalism looks far from scientifically respectable. Not only is the doctrine itself extremely nebulous and not obviously testable, it seems also to employ a highly questionable form of explanation. Since functionalism is rife in the social sciences, these problems raise serious doubts about naturalism.

Chapter 4 thus seeks to rebut this challenge. At the heart of functionalism are functional explanations. I peruse previous accounts of functional explanation and their weaknesses, providing in the process a basic sketch of functional explanations. Functional explanations are a species of causal explanation – and thus, qua explanations, perfectly acceptable. I then use my account of functional explanations to discuss in detail how they might be confirmed. Borrowing from tests for detecting natural selection, I sketch the kinds of evidence that can confirm functional explanations. I also point out numerous complexities previously overlooked; ignoring those complexities has led both to weak supporting evidence and simple-minded criticisms.

After arguing that functionalism can in principle be good science, I turn to empirical work – both good and bad. Functionalist social science often relies on optimality arguments, arguments that succeed only under stringent conditions. Yet those requirements frequently go unmet, a claim I demonstrate by looking at anthropological work on Inuit hunting practices and the Hindu beef aversion. However, not all functionalist accounts rely on such shaky evidence. In particular, I examine the recent work by Hannan and Freeman (1989) in organizational ecology. That work, I argue, produces well-confirmed functional explanations. Thus the moral of Chapter 4 is that functionalism can be not only rationally evaluated but confirmed. This is a striking blow for naturalism, since skeptics have long pointed to functionalist social sciences as a particularly clear example of pseudoscience.

Chapter 5 takes up the other main thesis of this book, namely, that a science of *society* is possible and necessary. Much good social science explains in terms of the large scale – in terms of institutions, organizations, groups, and so on. Individualists deny that these explanations are adequate. Thus my main defense of holism consists in rejecting individualist strictures on what the social sciences must be like. The first task in that argument is to distinguish carefully the various theses that are often run together by individualists and holists alike. Unless we do so, progress in the debate is impossible.

Individualists often claim that all good social theory can be reduced to a theory solely about individuals. Both individualists and holists try to determine the prospects for reduction on broad conceptual grounds. Neither side succeeds here, for the issue is ultimately an empirical one about the relation between theories, or so I argue. I thus go on to outline just what an actual reduction would involve and then, borrowing from work on reduction in the philosophy of mind and biology, I describe some general obstacles that potentially face reductionist claims in any domain. To defend the holist view, I need to show that those obstacles are real and persistent. Chapter 5 shows that they are by two routes: (1) by showing that paradigm cases of individualist theory do not provide reductions and (2) by identifying real cases of well-confirmed social theories that apparently are irreducible. Using the work of Becker and Downs as paradigm cases, I show that the rational choice approach provides little support for the reductionist program. I also show that the good social science defended in Chapters 3 and 4 – Paige's work on agrarian revolutions and Hannan and Freeman's work on organizational ecology – is irreducible, for the same general reasons that thwart reduction in other domains.

If individualism is implausible as a reductionist claim, it might nonetheless be reasonable as a claim about explanation or confirmation. Putting reducibility aside, could perhaps every social event still be explained in purely individual terms? Or perhaps individualist mechanisms are necessary to explain and confirm? Or maybe individualist accounts at least provide the best explanation? Despite their intuitive appeal, these claims are implausible, or so I argue. Social events cannot be adequately explained in terms of individuals, for individualist accounts will miss important patterns if reduction is impossible. Mechanisms are not necessary for explanation, nor are they essential to confirm macrosociological claims, at least as a general rule. And though individualists frequently claim that their theories are the best explanation, in many cases individualist and holist accounts need not compete, and when they do, holist theories can and sometimes do hold their own. The upshot of Chapter 5 is that individualism is seriously misguided. While there are truths in that doctrine – about the composition of society or about individualist detail in confirmation

Issues and arguments

and explanation – they are anemic truths, for even the most radical holist can grant them. When individualism is interesting, it is implausible; when it is plausible, it is uninteresting.

Chapter 6 confronts doubts about naturalism based on the meaningful nature of human behavior. To those from the interpretivist tradition, Chapters 3 and 4 will appear to beg the question. After all, I take social science evidence largely at face value. Interpretivists, however, will argue that the social sciences do not have the objective data nor produce the causal explanations typical of the natural sciences. Human behavior is essentially meaningful. Meaning, however, calls for interpretation and precludes a naturalistic science. Moreover, Quine and others in the analytic tradition reach similar conclusions, for they, too, worry that meanings are not suitable objects for ordinary scientific inquiry.

Aside from these general attacks, critics have also argued that many specific interpretive explanations are bogus. Social scientists frequently use norms and symbolic meanings to explain social behavior. Appeals to norms often seem to simply redescribe the fact to be explained – in short, not to explain at all. And social scientists seem to give norms an independent ontological status that would make even the most permissive cringe. Similarly, many anthropologists explain alien cultures by tracing out the meanings of rituals and other practices. Yet those meanings may be ones that no individual would acknowledge – and that is enough to cause some critics to see the whole enterprise as flawed (Skorupski 1976; Papineau 1978).

Chapter 6 thus takes up these challenges. I argue that belief-desire explanations need not be inherently circular, unrefinable, or indeterminate, and that intensional language can suffice for good science. Moreover, interpretive social science does not need some special source of evidence or special type of explanation, as a close look at interpretive work shows. Appeal to norms and implicit meanings can explain and be well confirmed, even if much work of this sort is weak. And even if these attacks on interpretive social science succeeded, they would be no serious blow to a science of *large-scale* social processes. Arguments from meaning, I shall suggest, rest on a strong and implausible individualist bias.

Of course, to show that an interpretive *science* is possible is not to show very much. After all, it might be only possible. Chapter 6 thus also looks at interpretive practice, focusing primarily on Brown and Harris's (1978) work on the social causes of depression. Producing good interpretive social science is hard, but Brown and Harris's work exemplifies the basic scientific virtues outlined in Chapter 2. Good interpretive social science is possible in practice.

Chapter 7 turns to economic theory. Economics is an apparent counterexample to both naturalism and holism. It raises doubts about naturalism because economics is seemingly the best of the social sciences and yet has had dubious empirical success. Much economic practice works out the mathematical subtleties of highly unrealistic models. Yet economists make little attempt to show that those models are confirmed and can explain, despite their implicit *ceteris paribus* clauses. Critics thus wonder whether economics should be treated as a science at all. Economics also raises doubts about holism because economics is seemingly our best social science and individualism is its official methodology. In the last decade economists have become even more aggressive about their individualism, claiming that no macroeconomics is adequate until it has individualist foundations. If individualism is plausible anywhere, it would seem to be in economics.

Chapter 7 shows in outline how to answer these claims. It is a mistake, I suggest, to treat economics as a monolithic whole exhausted by neo-classical general equilibrium theory. Most commentary on economics errs in this way. We need instead, I argue, to look in detail at the diversity of economic research if we want to evaluate what is good and bad in economic practice. Much actual economic investigation goes on independently of abstract neo-classical models. It works by applying what I call the supply-and-demand argument strategy on a case-by-case basis to identify causes; it thus proceeds much like the other good social and biological science discussed in earlier chapters. Some of this supply-and-demand work is relatively successful, or so I try to show.

I also try to show that individualism gains little support from economics. Microeconomics itself is frequently about social wholes, not individuals. Moreover, there is little prospect of reducing microeconomics, for all the standard obstacles to reduction are likely in the economic realm. Similar conclusions probably hold of macroeconomics as well. The common call for microfoundations is not warranted on any general methodological grounds; I try to use the machinery of Chapter 5 to say what sort of specific empirical circumstances might warrant a more restricted demand for mechanisms.

Chapter 8 concludes the book by putting the pieces together and using the earlier chapters to draw some general morals about the social sciences, their problems, and their prospects. I try to say why the social sciences have not fared better and where and how they may progress. Complexity is the usual excuse for poor social science. I emphasize other causes. Ideology often permeates social theory, much to its detriment. Social research is often not social enough – it lacks the collective yet competitive research practices of the natural sciences that promote such scientific virtues as objectivity and testability. These are contingent obstacles; a better social science can be had by eliminating them. Of course, such recommendations are no better than the concrete analyses and the picture of science on which they are based. It is to the latter that we turn in the next chapter as we take up challenges to scientific rationality.

2

Challenges to scientific rationality

Arguing for naturalism would be pointless if the natural sciences were not a paradigm of rational investigation. However, many philosophers and social scientists now argue that social forces drive the natural sciences. According to Kuhn, incommensurable paradigms dominate science, making scientific change a social process rather than a rational one driven by decisive tests. According to the social constructivists like Latour (1987), appeals to "evidence" and other epistemic values are really just a cover for negotiation and network building. These irrationalist views have enormous implications if true. Not only would naturalism be a non-starter, the social sciences would also have to be drastically reconceived. The drive for better data and more careful tests would be misguided. Much that social scientists do and struggle for would be without foundation. Moreover, social scientists would lose any claim to expertise on policy issues if their results are mere social constructs or one of many incommensurable paradigms. Social science would be just one more human conversation but with pretensions to be something it could not be.

This chapter tries to show that social scientists need not adopt these irrationalist doctrines. In what follows I try to do three basic things: answer the many irrationalist critics of natural science, identify what I take to be the basic core practices defining good science, and sketch the recent developments in philosophy of science that later chapters presuppose. The first two tasks are essential if we are to take naturalism seriously; the third, informational task will emerge in the process. (Readers familiar with recent philosophy of science or unconcerned with challenges to scientific rationality can skip all but Section 2.6 without detriment to later arguments.)

The argument proceeds as follows. Section 2.1 sets the stage by discussing Quine's (1953) path-breaking attack on positivism. Quine's criticisms made possible much of the currently trendy irrationalism, but his views also ground our best current answer to relativism. Section 2.2 clarifies issues. Numerous different theses are at work in debates over scientific rationality; discussing them undistinguished has caused considerable confusion. Section 2.3 takes up Kuhn's (1970) influential antirationalist views. He has convinced many social scientists that data cannot decide between theories, for data are theory laden and theories embody incommensurable paradigms. Section 2.3 explains why these claims are untenable.

Section 2.4 takes up more recent sociological approaches: social constructivism and Rorty's "post-modernist" pragmatism.[1] Social constructivists make the obvious, albeit long-ignored point that scientific institutions are social institutions. Accordingly, scientific beliefs must be formed by a social process. So, it is social negotiation, not evidence and the structure of reality, that explains scientific practices. Post-modernism presents an equally serious challenge. Rorty's claim that science is just another form of non-coercive persuasion has caught on among social scientists (for example, McCloskey 1985; Weintraub 1991). Section 2.4 examines these doctrines. We should, I argue, take the social nature of science seriously. But we can do so without committing ourselves to the dismal conclusion that all science is politics by other means.

Another threat to scientific rationality comes from the claim that science is inherently value laden. If normative elements constantly intrude in scientific practice, then we can hardly claim that science has any special lock on objective, impartial pursuit of the truth. Though there are important reminders here, I argue in 2.5 they are no serious threat to scientific rationality.

A general picture of good science will slowly emerge as we answer the irrationalists. Section 2.6 makes that picture more explicit. It also explains how we can say something useful about scientific adequacy while recognizing the contextual, contingent, and empirical nature of claims about scientific method. This last section thus more precisely formulates the initial naturalist theses identified in the last chapter. The general upshot is then twofold: first, we can defend the rationality of science without falling into simplistic positivist assumptions, and thus, second, naturalism in the social sciences is a live and fundamental issue.

2.1 **Quine and the demise of positivism**

Current philosophy of science owes its intellectual origins in large part to Quine. Quine's influence comes from the role he played in the

---

[1] Seminal works include Bloor (1976), Barnes (1982), Latour and Woolgar (1979), and Latour (1987). Rorty (1979, 1982b) are important works for the rhetoric of science.

downfall of logical positivism. The common thread to nearly all current philosophy of science is what it is not: positivist. And it was Quine's criticisms that led to the current post-positivist outlook. In this section I explain those criticisms and the basic framework for philosophy of science set by Quine.

Positivism, like other bygone intellectual movements, is probably inaccurately treated by its current detractors. Some or many positivists may not have held all the views currently falling under that rubric, and the positivists also changed their views over time (see Michael Friedman 1992). Nonetheless, the "positivist" program loosely construed has been an important factor in the way philosophers and social scientists have understood their work. The positivists believed that there are two kinds of truths: empirical, factual truths of observations and truths based on the meaning of words. Good science was the paradigm of the former; mathematics and logic, the paradigm of the latter. Like Hume before them, the positivists used this division to draw some striking conclusions. Traditional philosophy clearly was not based on facts of experience; yet, its metaphysical claims about substance, soul, and the like were not true simply by definition, and the positivists thus recommended – in Hume's words – that such doctrines should be "committed to the flames."

What then was philosophy supposed to do? Philosophers were not scientists and thus were not in the business of producing and testing empirical claims. Thus the positivists claimed that they were in the business of analyzing meanings (going bankrupt was not considered a live option). Philosophers should use their skills to analyze the concepts of science and common sense. Philosophy's job was to clean up conceptual messes.

Philosophy of science thus became primarily analyzing basic scientific concepts. Those concepts could either be theory specific – like the concept of mass in physics – or be metascientific notions like "is a law," "is confirmed," or "x is explained by y." So the positivists left much for philosophers of science to do, and what they left was something only philosophical analysis could provide. Analyzing concepts was not an empirical enterprise. For example, Carnap (1950, p. v) claimed that the concept of confirmation concerned "a logical relation between two statements" that "is not dependent on any synthetic statements" and is "analytic." Philosophy thus could give us something gotten nowhere else: the meaning and logic of science's fundamental concepts.

While the positivists were philosophical radicals in many ways, they were nonetheless quite traditional in others. Epistemology in its traditional form remained healthy in positivist practice. For example, some positivists were committed to "the given," the doctrine that sensory experience directly confronts us with information that is self-evident, relying upon no further inferences or theory. That idea had a long philosophical history and was far from radical. In the positivist's philosophy of science, the given appeared as "protocol sentences" or "observation reports" – the empirical bedrock of experience which is certain and from which all theories are derived and confirmed. Quoting Carnap (1934, p. 45) again, protocol sentences "refer to the given, and describe directly given experience"; they are "statements needing no justification and serving as the foundation for the remaining statements of science."

The positivists were traditional in yet another way. Although they ridiculed the idea that philosophers could produce a special kind of speculative knowledge, they did not practice what they preached. Under the guise of conceptual analysis, the positivists did not hesitate to rule on what constituted knowledge – on the requirements for justification, evidence, confirmation and other normative epistemological concepts. Moreover, since philosophy is not an empirical discipline, the positivists implicitly assumed that they had special methods that could distinguish good from bad science. However, this foundationalist enterprise was one Descartes, for example, could have been proud of.

There are, of course, numerous other ideas associated with positivism. But these core notions – that there is a clear distinction between empirical truths and conceptual ones, that science provides the former by beginning with the certainty of direct experience, that philosophy can, via conceptual analysis, tell us what explanation, evidence, and other scientific concepts require – formed a lasting legacy. That legacy has influenced how philosophers and scientists view the scientific enterprise and that legacy continues. Later chapters will find that unnoticed positivist assumptions still play an important part in debates over naturalism.

Positivism was officially put to rest in the 1950s. No doubt many factors were involved. The positivists themselves sometimes abandoned their most stringent views in the face of difficulties. However, the key intellectual factors were Quine's (1953) criticisms, for they sketched a broad philosophical framework that undergirds much contemporary philosophy of science. Quine's attack turned on challenging a crucial positivist distinction: that between truths based on observation and truths based on meanings. The argument worked on two fronts. Quine first argued that we have no clear notion of truth by definition or analyticity. Explaining analyticity in terms of synonymy, for example, goes nowhere. Synonymy is itself not a particularly clear notion and is most naturally defined by analyticity. Other attempts to clarify analyticity run into the same problem: either obscurity or circularity. Thus we seem to have no criteria for drawing a sharp distinction between analytic and synthetic truths.

Quine also raised a second, more systematic criticism. The analytic–

synthetic distinction tries to separate the linguistic and factual components behind our beliefs. Some statements are directly tied to confirming evidence or experience; they are synthetic. Other statements gain their credibility entirely from linguistic conventions and thus the empirical data can never refute them; they are accordingly analytic and *a priori*. Quine, however, denied that we could sharply divide evidence this way, because testing is a holistic affair. Following Duhem (1954), Quine argued that hypotheses do not confront experience or evidence one by one. Rather, testing a single hypothesis requires a host of background theory about the experimental apparatus, measurement theory, what data are relevant, what must be controlled for, and so on. So, when experiments fail, they only tell us something is wrong somewhere. We can save any hypothesis from doubt by changing our background assumptions. Theories face the test of evidence as wholes.

The moral is that evidence does not fall into two neat categories, one linguistic and the other empirical. We are faced, to use Quine's metaphor, with a "web of belief" (Quine and Ullian 1970). The evidence for any particular hypothesis depends on how well it fits with experience and our background theory. Individual observations that conflict with fundamental theoretical postulates lose credibility. Fundamental theoretical axioms that conflict with a host of empirical and theoretical data may be given up, even though earlier they looked to be true by definition. All parts of the web are indirectly relevant to all others. There is no absolute way to isolate the analytic, necessary truths from the merely empirical. In the end there are no *a priori* truths.

By denying a sharp conceptual–empirical distinction and pointing out the holistic nature of testing, Quine provided the intellectual foundations for a broad change in the philosophy of science. These two seemingly abstract philosophical ideas have wide-ranging ramifications, ramifications that are still being explored. Let me sketch those that matter most directly to the philosophy of science.

Since conceptual matters are not entirely distinct from empirical ones, philosophy of science can no longer be a purely conceptual enterprise. No doubt philosophers can still try to analyze confirmation, explanation, and the like. But those accounts ultimately will be empirical claims, not *a priori* conceptual truths. We thus have to rethink the relation between philosophy of science and scientific practice. Philosophers have no special place outside or prior to science itself. Foundationalism – as the idea that philosophers can describe on *a priori* grounds the standards for real scientific knowledge – has to go.

How then should philosophy of science proceed? Obviously it has to be tied to real scientific practice. That requirement is, however, quite weak.

Much current epistemology, for example, pays lip service to Quine's dictum. Yet it still tries to describe *a priori* conceptual truths about justification and other epistemological concepts (cf. Goldman 1986). Moreover, philosophy can be "tied to" scientific practice in different ways. Philosophers of science are still debating what the idea entails.[2] For Quine, philosophy of science must be "naturalized." Like the advocate of naturalism in the social sciences, the advocate of naturalized epistemology believes that human knowers are part of the natural world. They are thus best studied by the broad methods of the natural sciences. Epistemology must be an empirical discipline, one in the end that is continuous with science itself. This was Quine's message, though he did not always practice what he preached (see Chapter 6).

The idea that the philosophy of science must be continuous with science itself is widely accepted. Again, philosophers do not agree on what this stricture requires. They disagree both on what kinds of evidence are relevant to philosophy of science and on where that evidence is to be found. Some (Doppelt 1988) see philosophy proceeding roughly by the method Rawls described for political theory: by balancing intuitions (about good science) with philosophic accounts of scientific methodology until a "reflective equilibrium" is reached. Others (Larry Laudan 1990) see philosophy of science as largely making means–ends claims – judgments about methodology are judgments that particular practices best promote scientific goals. Philosophers also disagree on which empirical disciplines are most directly relevant. Quine thought naturalized epistemology was just a branch of psychology, the science that investigates our perceptual apparatus and thus our sources of knowledge. Cognitive science approaches to scientific reasoning are of a similar mind. Others see an essential role for the history of science, the sociology of science, or the numerous methodological studies inside the natural sciences themselves.

I shall not try to sort out these issues with any care. The naturalized philosophy of science employed here takes good methods to be those that promote scientific ends, above all knowledge or truth. However, making such judgments will inevitably rest on some prior sense about paradigm cases of good science; we cannot entirely avoid balancing intuitions and principles. It is also clear that psychology cannot tell us everything we want to know. The history of science and the sociology of science can help us see how science works and which scientific practices achieve their aims. Similarly, the methodological studies inside various natural sciences illustrate how to investigate empirically what promotes scientific goals. So,

---

[2] See, for example, Laudan (1984, 1987, 1990), Garber (1986), Doppelt (1990), Leplin (1990), Rosenberg (1990), Giere (1988), and Thagard (1988).

nineteenth century biology may help us evaluate reductionistic structures on good science; practical experience across numerous fields shows that double-blind experiments may be necessary to get reliable results. Thus construed, naturalized approaches do allow a place for normative evaluations – we are not stuck with simply describing what scientists do.[3] Although I find this general picture of naturalism most plausible, it is not essential: the symptoms of good science I outline below and use throughout the book do not presuppose any very specific version of naturalized philosophy of science.

Quine's naturalized epistemology suggests that the positivist ideal of a universal and substantive "logic of science" is misguided. The quest for such a logic has dominated twentieth century philosophy of science. Seeing why that quest is misplaced will help make it clear just how deep the Quinean criticisms go.

For the positivists, science was a rational and objective enterprise because it had a method. By method, they had in mind a set of universally valid rules much like those of formal logic. We can tell logically valid arguments simply by their form without attending to their content. *Modus ponens* (If P, then Q; P; therefore Q) holds for whatever sentences we plug in for P and Q. We know this inference is valid without making any specific empirical assumptions. The positivists thought that similar relations should hold between data and theory: given the data, we can apply the formal and universal rules of scientific inference to determine if the theory is confirmed. In short, the scientific method was the logic of science.

A logic of science thus has at least three basic requirements. It must be *universal*. Just as the laws of logic hold whatever the subject matter or time, so too good scientific inference is universal. A logic of science also requires that scientific method be *formal*; it should rest on no specific assumptions about the way the world is. Finally, a logic of science must be *sufficient*. Given a set of premises, rules of logic suffice to decide if a conclusion follows; we need no other information. A logic of science should allow us to do something similar – to decide whether a hypothesis is confirmed given the data. While the positivists did not always explicitly endorse these requirements, they are certainly implicit in their views.

A Quinean naturalized philosophy of science makes such a scientific method unlikely for several reasons. For one, any claims about the logic of science will be generalizations from scientific practice. Scientific prac-

---

[3] Critics also worry that naturalized epistemology rules out skepticism by fiat (see Stroud 1985). For my purposes, however, that problem can be put aside, since philosophical skepticism is a doctrine that will draw no wedge between the natural and social sciences.

tice is, however, diverse, both across time and across fields. That means finding *universal* rules will be difficult. Any rules that do describe all science must abstract from detail. However, abstract rules are unlikely to be *sufficient* because of the information they ignore. Moreover, the holistic nature of testing also makes sufficient rules unlikely. If testing is holistic, then we can apply criteria for good science only with the help of specific background assumptions. The more we abstract from concrete practice to devise universal standards, the more we will need domain-specific knowledge to apply those standards. Universal rules are thus unlikely to be sufficient.

Even if there is some set of scientific virtues that is universal and sufficient, they are unlikely to be *formal*. The positivists, remember, wanted formal rules that depend on no empirical assumptions about the world. If claims about good science are (1) ultimately empirical in nature and if (2) empirical evidence is holistic, then purely formal rules are unlikely. On the Quinean account, the rules of good scientific inference are just high-level empirical claims. What methods do promote scientific goals will depend on substantive facts about the way the world is, about the psychology of scientists, about the sociology of scientific institutions, and so on. Similarly, claims about good methodology will rely upon the relevant background assumptions. Those background assumptions will, however, be empirical in nature. Consequently, scientific methodology will rest on empirical assumptions.

Of course, the above reasoning does not *prove* that there is no logic of science. It does, however, show that the three criteria of formality, universality, and sufficiency are unlikely to be satisfied jointly. The more an account of method ignores empirical detail and domain-specific assumptions, the more abstract it becomes; the more abstract and simplified an account of method becomes, the less work it can do by itself in evaluating specific pieces of science.[4]

Classic attempts to find the logic of science have run into just the problems sketched above. I want to look briefly at why two classic accounts – hypothetical-deductivism and falsificationism – failed and how those failures have a natural Quinean diagnosis.

Hypothetical-deductivism (H-D) describes a familiar part of scientific practice. Here is how Richard Feynman (1965, p. 156) describes it:

In general we look for a new law by the following process. We guess it. Then we compute the consequences of the guess to see what would be implied if this law

---

[4] Siegel's (1985) defense of universal scientific method illustrates exactly this dilemma. For Siegel, the universal scientific method consists in "a commitment to evidence." While that seems right, it is so lacking in content as to be of little help in deciding substantive questions.

that we guessed is right. Then we compare the result of the computation with nature . . . to see if it works.

Carl Hempel (1965) has made admirable attempts to turn this simple idea into a sophisticated logic of science. The general consensus, however, is that the project has failed (Glymour 1980, Achinstein 1985). Suppose our hypothesis does predict what we subsequently observe. That agreement will constitute good evidence only if we know that there is not a more reasonable rival that also predicts what we observe. However, we can always construct rival hypotheses that fit the data. So agreement of hypothesis and observation is only convincing evidence if all rivals are inadequate. Furthermore, if a hypothesis H entails our data, then so does H conjoined with any other statement. Thus the bending of light around the sun would confirm the hypothesis "general relativity is true and Ronald Reagan is a transvestite." Of course, we know that the latter is irrelevant. However, hypothetical-deductivism gives us no formal way to rule out such cases. So the moral is that deducing observations is not good enough for confirmation – we have to build in a great deal of other empirical information, information that varies from context to context. Hypothetical-deductivism thus cannot be a purely formal account of confirmation.

Furthermore, *confirmation* is only one part of theory acceptance. Accepting a theory requires more than knowing whether a specific batch of data supports a particular hypothesis. It also requires that we factor in multiple tests, the scope of the data, the logical and evidential ties with other hypotheses, and so on. Even if H-D succeeded as a logic of testing, it would fall short of being the logic of science.

Popper's falsificationism fares no better. Hypothetical-deductivism fails because logically entailing the evidence is not enough to confirm. Yet, Popper argued, logically conflicting with the evidence is enough to *disconfirm,* for we need not know about competing hypotheses to know that the evidence rejects the hypothesis at hand. Of course Popper is right about this narrow point. Nonetheless, this superficial asymmetry between confirming and falsifying does not warrant the much more ambitious claims that Popper makes about methodology.

Three key elements in Popper's picture of scientific logic are the following claims:

(1) Falsifiability separates science from pseudoscience.
(2) The essence of scientific testing is the attempt at falsification.
(3) Good science is that which has survived attempts at falsification.

"Falsifiability" in the first claim requires that a particular hypothesis or theory be *logically* inconsistent with some potential data. Understood that way, falsifiability is a purely logical matter and thus is in principle suitable for a logic of science. However, falsifiability in this sense fails to distinguish science from pseudoscience. Hypotheses *by themselves* will generally be neither consistent nor inconsistent with the evidence. Only by bringing in what we might call a "theory of the test" will the hypothesis conflict or cohere with the data. A theory of the test includes all those background or auxiliary assumptions that tie hypotheses to data. Assumptions about what the data measure, about potential confounding factors, about the experimental setup, and so on are essential before the data tell us anything about theory. Individual hypotheses are seldom falsifiable.

Of course, we can shift from the falsifiability of single hypotheses to the falsifiability of a hypothesis plus auxiliary assumptions. However, this modified criterion will not give us what Popper wants for three reasons. First, we can make any piece of nonsense falsifiable if we join it with the suitable background assumptions. For any hypothesis H, we can just add the further assumption that H entails P, where P is a description of some data. H will then be falsifiable. Surely this violates the *spirit* of Popper's criterion. However, he has no purely formal way to rule it out, as should be the case if falsifiability is part of the logic of science. Second, the broader falsifiability criterion no longer separates falsification from confirmation, though Popper thought confirmatory evidence totally inadequate. A hypothesis that entails the data is well confirmed if we know that no other hypothesis does so. For example, consider the symptoms that are the differential diagnosis for a disease. If they really are differential symptoms, then they are compatible with only one diagnosis. So a diagnosis that is consistent with those symptoms plus the background knowledge that no other diagnosis is consistent conclusively confirms the diagnosis. So falsification and disconfirmation turn out to be two sides of the same coin. Finally, allowing in the theory of test means that falsifiability is no longer a purely logical matter. Rather, to determine how data bears on a particular hypothesis, we have to bring in substantive, domain-specific background knowledge. This means falsifiability is no longer a formal criterion and thus does not give us a logic of science, however useful it may otherwise be.

Once we see the difficulties with falsifiability, we can also see that the other components of Popper's methodology – testing as attempts at falsification and surviving such attempts as a criterion for good science – are no logic of science either. A good test is not simply one where the hypothesis had a chance of being disconfirmed, nor is good science simply that which survives attempts at falsification. Failing a test sometimes reflects on the theory of the test, not the hypothesis at issue. And scientists on occasion quite reasonably ignore failed tests for just this reason. Moreover, tests with controls – the prime way we ensure a chance of disconfirmation – can be lousy tests, if the test does not rule out competing explanations. For example, the hypothesis that oat bran lowers serum cholesterol initially was tested by comparing those who ate oat bran with those who ate a regular diet. Yet that test was a poor test, for it did not eliminate a plausible competing hypothesis, namely, that eating large quantities of grain meant eating lower quantities of fatty food. So searching for falsification provides neither a sufficient criterion nor one that works independently of substantive empirical background knowledge. As a logic of science, Popper's falsificationism fails.

We predicted on Quinean grounds that there would be no logic of science. The empirical nature of methodological claims and the holism of testing would combine to make any informative formal criteria of good science unlikely. That is exactly what we have found. Neither deductivism nor falsificationism provides a *substantive* logic of science, for each depends extensively on contingent, contextual, and substantive empirical assumptions for their content.[5] The positivists wanted much more.

However, the moral of our Quinean story is not simply destructive. Surely hypothetical-deductivism and falsificationism get at something important. For example, the search for sharp, falsifiable hypotheses no doubt plays an important role in good science as does the process of comparing implications to data. However, if there is no simple logic of science and if ultimately claims about scientific methods are empirical claims, we cannot evaluate social science by looking at simple formal traits. We must instead look in detail at the various different claims, kinds of evidence, and concrete methods that are used in the social sciences. It is those factors that make or break a scientific enterprise, and it is on those grounds that the

---

[5] The Bayesian approach can provide a very useful framework for thinking about methodological issues, as is ably demonstrated by Howson and Urbach (1989). However, it seems clear that Bayesian approaches do not provide the kind of logic of science that philosophers have wanted, for they are entirely parasitic on prior information about important methodological questions. For example, Bayes's theorem is compatible with entirely different views on such fundamental normative questions as what constitutes a good explanation, what role explanatory information plays in confirmation, whether data used in constructing a hypothesis count less than novel data, whether simplicity has probative force, and so on. In each of these cases, we could have empirical information that led to one answer or the other, and Bayes's theorem must await the result of those investigations, e.g. we might discover that novel data indicate a reliable process and thus count for more (see Maher 1988), that particular versions of simplicity do up the odds of truth, that appeals to explanatory power generally are less reliable than other empirical virtues, and that specific constraints on explanation are incompatible with our best theory of the world. All of these results could be plugged into Bayes's theorem, but only after we had decided these fundamental issues.

prospects for a science of society must be decided. The Quinean revolution tells us that there is no real alternative.

So the Quinean transformation implies much, for both how and how not to do philosophy of science. Though the implications I have outlined seem far-reaching, some think even more radical conclusions follow. The most serious challengers to scientific rationality also argue from Quinean premises. It is to these challenges that I turn next.

2.2. Varieties of rationality

Kuhn, Rorty, and the advocates of social constructivism hold what we might call an irrationalist view of science; they deny that science has any particular claim on rationality. Defenders of naturalism, on the other hand, want to show that the social sciences can be good science precisely because the natural sciences are the paradigm of a rational investigation. Yet, "rationality" is a notoriously vague word; debates over scientific rationality are often unproductive, because it is not clear just what theses are at stake. Thus, before we can consider the arguments on either side, we need to clarify the issues. Below I distinguish multiple aspects of scientific rationality and use them to identify several different rationalist and irrationalist positions.[6]

When someone claims that science is rational, they may be making a claim about

> *current science versus science throughout history.* Arguing that past science is fully rational is clearly a much stronger thesis than that "only" current science is.
>
> *scientific change versus stasis.* During periods of scientific change, standards may be in dispute. Consequently, irrationalists may grant that ordinary science is rational but deny that transitions are.
>
> *science as evaluated by current standards versus as evaluated by standards at the time.* On the simplest, most naive view there has been one homogeneous activity called science that began roughly with Galileo – since then science has progressed by applying the scientific method. Such a claim is much stronger than a view that allows scientific standards to change.
>
> *evidence construed narrowly as primarily empirical adequacy versus evidence broadly construed as "good reasons."* Consistency with empirical data is the traditional empiricist requirement for rational scientific belief. Critics deny that empirical adequacy

---

[6] Helpful work towards this end is found in Hellman (1983) and Doppelt (1988).

plays any such role in science; they allege that simplicity, consistency with metaphysical world views, cultural norms, and so on equally motivate the adoption of scientific theories.

the reasons actually used by scientists versus the reasons that could have been given at the time. Scientists may sometimes come to believe their theories for non-rational reasons, even though they could have reached those same views from evidence available at the time.

reasons that uniquely determine which theory to accept versus reasons that are relevant to a theory. The reasons for a scientific theory may be reasons in the sense that they favor only that theory; alternatively, scientists may adopt a theory for good reasons, though those reasons did not uniquely warrant that choice.

rationality as having reasons or evidence versus rationality as truth. Scientific realists want science to be rational in the sense that it is true, approximately true, or getting truer. They obviously want more than those who claim that current science has good evidence but nonetheless may not be true.

These dimensions do not exhaust the possible contrasts; they certainly illustrate just how complex questions about scientific rationality are. For our purposes, we need not specify all possible views in this conceptual space. Instead, let me distinguish some major positions according to how far they are willing to advance down the following list of theses:

(1) Current science has good evidence by current standards.
(2) Past science has good evidence by past standards.
(3) Current science developed from past science because of good evidence by standards shared at the time.
(4) The standards mentioned in (1) through (3) are identical, that is, there are universal standards of good science.
(5) Evidence in (a) current or (b) current and past science is primarily empirical adequacy.
(6) Current science is approximately true.

The first thesis makes a claim only about current science. Without some further constraints on what constitutes evidence, it is a quite weak claim. Theses (2) and (3) make assertions about the rationality of past science and scientific change, respectively. Thesis (4) asserts not just that current scientific norms arose through some rational process but that those standards hold for all times and places. The fifth claim requires that "evidence" be interpreted in the narrower sense of observational adequacy. When added to the other five, the sixth thesis requires not just that science be a rational process but that it succeed in getting the truth. In short, it ties rationality to the doctrine of scientific realism. That transition is one opponents of realism like van Fraassen (1981) would reject.

Radical antirationalists deny all these claims. This, I shall argue, is really the position of the social constructivists: its advocates deny that evidence as traditionally understood plays any decisive role, either in scientific change or stasis, by current standards or past. Social causes, not reasons, underlie science. Moderate antirationalists like Kuhn accept the first two theses but go no further. Kuhn denies that shared standards ground scientific change. He likewise denies that empirical adequacy plays a primary role in scientific change[7] and denies that science is getting closer to the truth. Radical rationalists of course want it all: not just good empirical evidence evaluated by shared standards but also truth. Kitcher (1993), for example, argues for this position.

To defend naturalism I obviously need to argue that the natural sciences are a paradigm of rational investigation. However, I do not need to defend a strong scientific realism to do so. We might never know that current science is largely true even though it has good evidence. Nonetheless, natural science could still seem our best shot at the truth, given the evidence we have. Consequently, naturalism about the social sciences would remain a highly significant and contentious issue – even if scientific realism were implausible. So a significant naturalism need not affirm thesis (6).

What then does a more moderate rationalism claim? Multiple versions are possible. "Moderate rationalism" would require theses (1) and (5) only. In short, it would demand that current natural science have good evidence by current standards, with evidence defined as primarily empirical adequacy. Without this latter restriction, rationalism would let nearly everything in, since even astrology and deconstructive literary criticism provide evidence in some sense. A stronger position thus holds not only (1) and (5) but also defends the rationality of scientific change – in short, some variant of (2) and (3).

In the rest of this chapter I shall straddle the fence between moderate

---

[7] If this attribution seems unfair, consider the following: "Given a paradigm, interpretation of data is central to the enterprise that explores it. But that interpretive enterprise . . . can only articulate a paradigm, not correct it. Paradigms are not corrigible by normal science at all. . . . [Transition to a new paradigm is determined] not by deliberation and interpretation [of data], but by a relatively sudden and unstructured event like the gestalt switch" (Kuhn 1970, p. 122); the relative problem-solving ability of a paradigm is "neither individually or collectively compelling" (p. 155).

and strong rationalism. Research in the sociology of knowledge has not shown scientific evidence to be a mere social construct, nor has Kuhn shown that scientific standards do not change for rational reasons. So strong rationalism (in some form) is a potentially plausible position. Showing more than that is far too great a task for this book. Fortunately, I can straddle the fence in this way without interfering with my main task – defending naturalism about the social sciences. For, if by our best lights current science is a paradigm of rationality, then naturalism is an important issue no matter how we came to practice the science that we do. Moreover, I shall at least undercut irrationalist skepticism about those origins.

2.3 Kuhn and shifting standards

Thomas Kuhn took Quine seriously and looked empirically at the sciences. He found numerous phenomena that led him to reject the rationalist's position. Kuhn's main arguments turn on three basic theses: the incommensurability of meaning and standards, the theory-laden nature of data, and the malleable nature of scientific standards. Between them, these theses still constitute the most sophisticated challenge to rationalism. I want to discuss these rationales one by one. (Throughout my concern is with the Kuhn found in the original edition of *The Structure of Scientific Revolutions;* the later Kuhn [1970, Postscript] seems to have backed away from the original radical implications and is not my subject here.)

2.3.1 *Incommensurability*

Kuhn's most radical claim is that different paradigms cannot communicate. His basic argument is this: The meaning of terms in a theory is determined by their role in that theory – by their relations to other terms, the assertions of which they are a part, and so on. However, scientific change is revolutionary; the transitions from Aristotle to Newton and Newton to Einstein, for example, were radical changes in theory. Thus Newtonian "mass" and relativistic "mass" must have entirely different meanings, since they are parts of very different theories and theories determine meaning. In short, we cannot translate across paradigms.

Kuhn's argument is implausible. It both has absurd consequences and presupposes an implausible theory of meaning and translation. The absurdity comes in two forms. If different paradigms speak in entirely different languages, then they really never disagree. Since they share no meanings, they cannot assert what the other denies (Shapere 1964). Moreover, if meaning depends entirely on the overarching theory, then every difference in theory produces differences in meaning. So when any two individuals have different beliefs about the world, meanings will differ as well. According to Kuhn, however, differences in meaning preclude successful translation. Those who do not share Kuhn's theory of science should be unable to understand him!

These absurdities point to a deeper problem confronting Kuhn. Kuhn's arguments do not result from a well-confirmed theory of meaning and translation. No one has such a theory, and Kuhn's armchair account is unpromising for two reasons: it ignores the apparent fact that meaning dependence is not an all-or-nothing matter and that sameness of *reference* rather than sense may suffice for translation.

A term or sentence's meaning apparently depends more on some connections than on others. If meaning is holistic in this more limited way, then not every change in theory is a change in meaning and changing the meaning of one term need not change the meaning of all. Room is thus left for different theories to share meanings. Newton and Einstein may have used "mass" differently, but not so differently that no plausible translation is possible. Even if we could not translate "mass," that hardly means that Newton and Einstein used "telescope," "material body," "planet," and other more observational terms in radically different ways. So data can be shared and theories tested. Kuhn can preclude this outcome, but only by arguing that any change in theory entails a radical change in meaning. Thus either translation is possible or Kuhn is stuck with the absurdities mentioned above.

Even if meanings do vary significantly across theories, incommensurability is not inevitable. For purposes of theory comparison, the important connections may be ones of *reference,* not meaning. Reference does not exhaust meaning, because terms with different meanings can have the same reference. Since that is the case, meaning may vary across theories without preventing one theory from seeing what objects the other theory is talking about. Kitcher (1978, 1993), among others, has defended this route around incommensurability. He has argued that we can identify the referent of "phlogiston," for example, in terms of current chemical theory. Doing so will be a complex process, for "phlogiston" may not have a constant reference in all contexts. However, once we determine the referent of "phlogiston" in the relevant contexts, then we can understand and rationally evaluate earlier chemical theories. Similar approaches have been taken to "mass," Kuhn's other favorite example of an incommensurable term (Pearce 1987). Hence meaning variance need not prevent translation.

Kuhn has another, more plausible reason for thinking that theories are incommensurable: they invoke different standards for good science. At various times infallible knowledge, inductivist methodology, mechanical explanations, and mathematical tractability, for example, have been sine qua nons of good science. If, however, such basic standards change with changes in theory, then disputes between paradigms cannot be settled rationally. Each theory will appeal to its own conception of adequacy and no rational winner is possible.

Shifting standards do not make theory choice non-rational. Kuhn errs by treating theories as monolithic, undifferentiated wholes. More specifically, he thinks scientific change involves every aspect of science – goals, methods, standards, data, and theory. In analyzing historical change, this assumption means Kuhn "telescopes" the transition from one theory to another. If we take wide enough spans of time – say Aristotelian-Ptolemaic physics at one end and nineteenth century Newtonian physics at the other – and compare theories, then we may indeed see changes in every part of scientific practice. To see scientific change this way, however, is to collapse the historical process and ignore the details of change. Long-term changes may be radical. Yet the short-term changes that realize them may be only piecemeal. If change is piecemeal, we can use one part of a scientific practice to motivate change in others. In particular, we can rationally argue for change if we share other goals, standards, data, and theory. We can then learn that some standards do not promote our cognitive goals or are inconsistent with our best theories of the world. Looking only at the endpoints of scientific change, we miss this process. Newtonian physics, for example, started with the demand that science proceed only inductively. By the nineteenth century, physicists saw that such a stricture fit poorly with their best theories. Something similar happened with the early mechanical philosophy when it required science to eschew action at a distance and provide infallible knowledge.

While the prospects for piecemeal change are thus not mere idle speculation, I obviously cannot prove my case here. Doing so would require detailed historical studies. Much such work has been done by others.[8] Equally important, Kuhn himself does not establish that his favorite examples of paradigm shifts were entirely monolithic. Instead, he simply focuses on the endpoints of long historical changes and argues that standards, methods and theories have all changed. Such evidence is entirely inconclusive. Since my argument only presupposes that current science is a paradigm of rationality, this conclusion will do. Kuhn has not demonstrated that scientific change is irrational because of shifting standards, and there is some good evidence to the contrary. On the other hand, I cannot accept the claim that choice between *current* theories is non-rational because of diverse standards. I will take up that issue in analyzing

[8] See, for example, Gallison (1987), Zahar (1989), R. Laudan (1981), Franklin (1990), and Kitcher (1993).

another Kuhnian claim – namely, that theory choice rests on theory-laden data.

2.3.2 *Theory-laden data*

Data are theory laden when the data used to test a theory presuppose the very theory at issue. However, if the theory at issue determines the data, then we should worry (a) that every test of the theory is circular, since it presupposes its own truth in the process of testing, and (b) that competing theories will determine different data and therefore no data will be able to decide between them. Rational adjudication looks impossible.

Kuhn has several reasons for thinking that theory infects data. One rationale we have already seen and criticized, namely, meaning incommensurability. Other reasons turn on the fact that different theories ask different questions and on the holism of testing. Although Kuhn is surely right about these facts, they do not warrant irrationalist conclusions.

Different theories surely do ask different questions. Yet no drastic conclusions follow. After all, geology and physiology ask different questions as well, yet that is hardly reason for thinking that all testing is ultimately circular. Asking different questions does not preclude answering those questions with neutral data according to shared standards. Moreover, theories that differ on some questions may agree on others. Even though Aristotelians, for example, asked why motion continued and Galileo asked why it changed, there were other questions and evidence they shared. Galileo and subsequent physicists showed that focusing on change in motion was the best route to answering other questions that physicists did share. Finally, great theoretical differences – differences in "paradigms" – do not always mean great differences in the questions asked. For example, physicists in the 1960s debated a fundamental tenet of modern physics: general relativity's claim that the gravitational constant is invariant (see Will 1986). The Brans-Dicke theory, the competitor, eventually lost out after an intense debate and experimental results. Though they differed on fundamental theory, the competitors did share a broad framework of questions, enough to convince Brans and Dicke to give up their theory.

Kuhn's last argument for theory-laden data is probably his strongest. Testing, as Duhem and Quine taught us, is holistic. Every experiment requires background theory – what we earlier called "the theory of the test" – to interpret the experimental situation. But where does the background theory come from? Kuhn thought it came from the very paradigm being tested. Thus both in obtaining data and in determining what the data entails, we use the theory at issue. So testing is ultimately circular and relies on theory not shared by competitors. Data cannot rationally decide between theories.

Kuhn infers from the holistic nature of testing to the claim that every test is a test of the whole. That conclusion does not follow. The basic difficulty is one we have already seen: treating theories as monolithic blocks, ignoring internal complexity. "The theory" or "the paradigm" gives a single name to what is really a diverse batch of claims, methods, goals, and practices. Although every hypothesis relies on background theory, the theory of the experiment need not be the same as the theory being tested. Because theories or paradigms are composed of many different claims and methods, we can use one part to test another. If the theory we are testing is not used to determine the data, then no circularity threatens.

How do we tell if a theory is being used to test itself? Roughly the idea is this. The background theory and the hypothesis at issue should be (1) logically and (2) evidentially independent. If the presupposed background theory entails the hypothesis under scrutiny, then obviously the test is circular. Likewise, if the evidence for the hypothesis is the same as the evidence for the theory of the test, then we are testing only the whole. However, when conditions (1) and (2) hold, we can attribute blame or credit to just the hypothesis at issue.[9]

Let me discuss several examples that illustrate the arguments I have just made. A first example comes from electron micrographs (EMs). EMs are a central tool for testing hypotheses in cell biology. Of course, EMs rely extensively on background theory. Yet, this powerful tool depends on chemistry and physics, not on assumptions about cell biology itself. Physics tells us about the wavelength of electrons and how they scatter in different materials. Chemistry tells us about the reactions involved in staining samples. Not by the furthest stretch of the imagination, however, does quantum mechanics or the theory of chemical bonding rely essentially on cell biology for its evidence. So if these tests in cell biology depend on background theory, it is an independent theory with independent evidence.

A second example, due to Franklin (1984), comes from possible tests of Newtonian versus relativistic physics. Imagine a test of these two theories that involves creating a collision between two billiard balls with known velocity. After the collision, we measure the angle of scattering and the resultant velocities. Newtonian physics predicts the angle will be 90 degrees while relativistic physics does not. While the two theories will calculate the initial masses and momenta differently, they share an important range of background theory. Both accept the obvious sensory information – a ball is observed, balls collide, and so on – as well as the initial

---

[9] These ideas have been developed in much more careful detail by Kosso (1989). See also Glymour (1980).

velocity and the manner of measuring the scattering angle. Since the two theories make different predictions about these elements, data can decide between them. This example, like that of EMs, shows in practice what I defended above in theory: the holistic nature of testing does not entail that data is theory laden in any objectionable way.

2.3.3 *Ambiguous criteria*

So far we have answered Kuhnian challenges to scientific rationality based on the incommensurability of meanings, standards, and questions and on the theory-ladenness of data. I want now to consider one last irrationalist argument from Kuhn: the idea that criteria of theory choice are inherently ambiguous.

In an essay after *The Structure of Scientific Revolutions*, Kuhn (1974b) identifies five scientific virtues: accuracy, consistency, scope, simplicity, and fruitfulness. Though Kuhn believes these criteria are common to all or most science, he denies that they suffice for deciding between competing theories. According to Kuhn, "two men committed to the same list of criteria for choice may nonetheless reach different conclusions" (1974b, p. 324). They can do so because they can either interpret particular criteria differently or because they give individual criteria different weights. Simplicity, for example, can be read in many ways. Moreover, accuracy may be gained at the expense of simplicity, but there is no algorithmic way to make such tradeoffs. As a result, even shared scientific standards will not settle debates across paradigms.

Basic scientific virtues surely do admit of multiple interpretations. Here Kuhn has the holism of testing on his side. Claims about good science are empirical claims, ones that depend on our background knowledge for their interpretation and justification. Empirical adequacy, for example, may be shared in the abstract and yet mean different things for theories with very different views of the world – just as a slave holder and an abolitionist can agree on equality as a moral virtue and yet disagree on slavery, given different empirical assumptions about human nature. Kuhn is also surely right that competing theories can succeed and fail on different criteria, leaving the two at a standoff.

Nonetheless, Kuhn's irrationalist conclusions do not follow from these useful points. Consider first theories that interpret fundamental scientific virtues differently. The question is again whether these theories share enough other background theory to resolve the dispute, just as it was in the cases of shifting standards and theory-laden data. For example, Ptolemaic astronomers agreed with Galileo and Kepler that astronomy should save the phenomena; they had different ideas about how to achieve that goal. Ultimately, their rejection of the telescope lost out. For a Kuhnian argument to succeed, it is not enough to show that these two schools understood empirical adequacy differently. It would also have to show that they did not share sufficient background to decide, after debate and investigation, whether the telescope provided reliable information.

Different theories can also succeed on different criteria without making theory choice a "conversion experience." Competing theories frequently have different virtues. The rational response is to look for theoretical developments and further data that will decide the issue – by showing, among other things, that one hypothesis is the best *no matter what criterion or data set is emphasized.* For example, in the recent debate over continental drift, defenders and critics of that theory arguably emphasized different criteria of scientific adequacy. Supporters pointed to a variety of confirming evidence and critics demanded novel predictions. Eventually, drift theory won out because it satisfied both criteria (Laudan and Laudan 1989). So divergent interpretations do not entail irresolvable disagreement.

We have now surveyed and rejected Kuhn's radical arguments. At most he shows that irrational factors are possible. We can sketch apparently standard scientific practices that suggest that such problems are not actual. As an attack on rationalism, Kuhn's views are unconvincing. However, leaving our assessment here would be to ignore Kuhn's many subtle insights into how science works. I want to end this section by sketching some of them.

The most obvious insight is that testing is holistic. Of course, Duhem and then Quine made this point earlier, but not with the kind of depth and scope found in Kuhn. He shows how standards of good science, metaphysical world views, interpretations of scientific virtues, skills, research strategies, and much more are involved in tying hypotheses to the world. Kuhn thus gives the idea that testing relies on "auxiliary assumptions" much more meat. Kuhn also shows that science involves much more than theories conceived of as sets of statements. Heuristics, problem-solving strategies, skills, implicit knowledge, and the like are part and parcel of scientific practice. Conversely, theories and models serve many roles other than describing reality. They are embedded in scientific practice; they thus also perform many non-explanatory and non-predictive functions.

Another Kuhnian lesson is that theories are ephemeral entities considered in isolation from the concrete investigations they motivate; we must understand paradigms via the concrete practices embodying them. So "the" paradigm or "the" methods characteristic of a particular science are a gross simplification; their real content comes in their application. Applications, of course, vary, so theories and methods will take on different readings and different meanings in different contexts.[10] This means that we must look at the actual practice of science rather than just its textbook formulations.

Finally, Kuhn's incommensurability doctrine does have a useful if less profound reading – namely, that different theories need not divide the world in the same way nor even divide them in ways that can be systematically matched onto each other. This is one way to understand the idea that scientists "live in different worlds" without extreme idealist implications (see Hacking 1993).

Of course these ideas are vague as they stand. Nonetheless, we will make good use of them when we discuss the symptoms of good science, causal explanation, the role of models in economics, the relation between macro- and microsociological kinds or descriptions, and other topics later in the book. They are useful antidotes to lurking positivist assumptions.

2.4 Social constructivism and post-modernist rhetoric

While Kuhn thought science was essentially a social process, he did not completely abolish the idea of evidence and data. Some of Kuhn's successors have not hesitated to take that further step. On their view, scientific belief, like all belief, is caused by natural processes, not by such mysterious entities as "the data" or "the scientific method." The only difference between science and any other activity is a sociological one.

In this section I look at some recent work from two approaches: (1) the broad movement I have labelled "social constructivism" and (2) an approach emphasizing the rhetoric of science. Advocates of constructivism and the rhetorical approach do not always agree among themselves. I will not attempt to trace all the ins and outs in these two movements. Instead, I shall discuss the work of Bloor and Latour on social constructivism and of Rorty on science as persuasion. These are seminal figures for these approaches. Seeing where they go wrong will give us a good basis to judge both positions.

Social constructivism comes from social scientists dissatisfied with past sociology of science. Traditionally, historians and sociologists of science had distinguished internal from external factors. Internal factors concerned the data, methods, reasons, and theories present in a particular scientific period. These factors were rational ones. The social scientists first sought to explain scientific belief internally. When such explanations failed, the historian then invoked external factors. External factors were generally the social causes of belief – for example, racial prejudice or theo-

---

[10] For a further development of this idea, see Cartwright (1993).

logical assumptions. These external factors were non-rational and the explanation of last resort.

Constructivists like Bloor (1976) find this distinction odious. As sociologists of science, their task is to explain science scientifically. Looked at scientifically, however, beliefs are natural objects in the universe; *all* beliefs are caused, even those that we normally think of as motivated by the evidence. Bloor, for example, calls this the "causality principle" and makes it a central tenet. Scientific beliefs are to be explained by social processes; they are not somehow outside the causal network and produced by such abstract entities as "the data" and "the scientific method."

Not surprisingly, Bloor also rejects the idea that there are two kinds of factors in science, the rational and non-rational. We may identify some beliefs as true and others as false, but "the same type of cause would explain both" (1976, p. 5). The traditional history and sociology of science, Bloor thinks, assumed just the opposite – namely, that social factors explain the unjustified beliefs of scientists and non-social rational factors explain the rest. Bloor rejects this asymmetrical treatment, advocating instead the "symmetry principle": all scientific belief has the same type of cause. Rationality falls by the way.

Social constructivism mingles obviously justified complaints about past sociology of science with completely unjustified inferences about scientific rationality. Surely science is a social process and scientific belief does have social causes. Yet that fact does not *entail* that evidence, reasons, scientific method, and rationality do not ground science, for several reasons.

For at least one sense of scientific rationality, the causal origin of a belief is irrelevant. In Section 2.2 we distinguished between reasons that we can give for a theory versus reasons that actually lead to its adoption. Even if scientists have their beliefs for social reasons, those beliefs might nonetheless have good evidence. For example, if I have a dream that causes me to believe arithmetic is incomplete, that does not preclude there being a proof for the claim as well. Or, racism and political agendas may explain why IQ studies are interpreted as they are, yet we can still ask whether those interpretations are valid. Beliefs can be caused and still be reasonable. Social constructivists implicitly and explicitly make these judgments regularly, even though they are strictly speaking not entitled to them.

More important, reasons can be causes.[11] If reasons can be causes, then scientific belief can be caused, and be caused by a social process, and still result from evidence and scientific method. A rational process causes scientific beliefs if that process works through beliefs about the data and

---

[11] The classic defense of this claim is Davidson (1980).

about the best methods for pursuing the truth. This process can likewise be thoroughly social. Beliefs about the evidence and the best methods can come from interacting with others. So long as that process results from evidential beliefs, then social causation and rationality are not necessarily incompatible. Prestige and pecuniary gain may motivate scientists. Yet if the scientific community rewards the search for evidence and good theories with such commodities, then beliefs about evidence and so on can still drive scientific practice.[12] Science will not be a mere social process.[13]

So Bloor's symmetry principle – namely, that rational and irrational beliefs have the same kind of cause – is either compatible with scientific rationality or begs the question. If the principle covers any kind of cause, including beliefs about evidence, then rationalists will not disagree. If the principle requires "the same kind of cause" to be a non-rational cause, then the principle simply begs the question. For, it now says we are to explain all scientific beliefs in terms of *non-evidential* reasons. That assumption, of course, is just the question at issue.

Finally, even if social processes that do not work through evidence cause scientific beliefs, those causes might still promote rationality.[14] Suppose that in disputes over evidence we find that the most powerful scientists always win. It might nonetheless still be the case that those in power are generally *right* – that the social hierarchy on the whole plays a positive epistemic role. Naturalized philosophy of science evaluates methods by how well they promote scientific goals. So if scientific beliefs are produced by a social process, we can still ask how well that process promotes truth.[15] Or, to take a more accessible goal, we can ask with what frequency meth-

---

[12] For an abstract model showing how pursuit of gain can nonetheless lead to truth promotion, see Goldman and Shaked (1991).

[13] Social constructivists of course do not like the adjective "mere" here, though it is not at all clear how they can deny it is appropriate and still maintain the radical relativism of the approach. When Andy Pickering, for example, explains why the description is inappropriate, he in effect backs away from the radical versions of the approach and allows that "reasoned static appraisals of the relations between theory and evidence [are] vital to science" (1990). Pickering denies that those appraisals are sufficient to decide scientific controversy and thus his view is much akin to that of Kuhn on ambiguous criteria.

[14] This latter possibility is developed by Goldman (1987).

[15] Of course this presupposes some independent evidence about what the truth is. However, that assumption only looks troublesome when we focus on the rationality of theories as a monolithic unit or if we are defending rationalism in the strong form that defends the claim that we know current science is largely true. Once we recognize the diversity of claims, processes, types of evidence, and so on that make up real science, the claim of circularity is much less obvious – as our discussion of Kuhn pointed out.

ods promote empirical success. The fact that science is a social process does not tell us the answer to that question.

This third alternative thus suggests that science as a collective activity can produce rational results even if its component practitioners are less than fully rational. In short, it may be that the scientific community as a whole produces rational outcomes even though individual beliefs are not based on a correct and careful assessment of the evidence and so on. Science is a collective process, and many traditional scientific virtues such as objectivity are perhaps best thought of as properties of the community as a whole. For example, biases pulling in opposite directions may produce a collectively unbiased outcome. In short, a rational scientific community need not be a community of rational scientists. Both social constructivists and defenders of rationalism generally miss this possibility, I suspect, because of hidden individualist assumptions (see Solomon 1994 for an important exception).

So beliefs *can* be caused without thereby being irrational. A full rationalist defense would involve weaving at least the three possibilities sketched above into a story about how real science works.[16] That story would no doubt be complex, but pointing out that science is a social product in no way shows such a story impossible. Thus the pessimistic conclusions of the social constructivists are at best an empirical possibility. Obviously I cannot assess here the empirical merits of every study allegedly supporting constructivism.[17] However, I do want to look briefly at some of the most interesting constructivist work, that of Latour.

Latour argues that the process of science is a social process of negotiation, one in which actors seek to build bigger and bigger networks. Establishing scientific fact is establishing an unassailable network. Appeals to evidence, rationality, and nature are really strategies for defending that network; they are honorific terms for the process of social negotiation that constructs science. In *Laboratory Life* Latour along with co-author Woolgar (1979) studied the "discovery" of TRH, a pituitary hormone. They concluded that there was no logic of science that dictated when TRH was found; it was social negotiation that turned hypothesis into fact. In *Science in Action* Latour (1987) describes how scientific networks are built using scientific literature, laboratories, machines, and experiments. Facts, evidence, and rationality drop out and are replaced by network building.

Latour's work is often very insightful, particularly when it comes to describing the minutiae of scientific practice. Yet its social constructivist

---

[16] See Kitcher (1993) for a start on such an account.
[17] See Cole (1992), J. Brown (1989), Gallison (1987), and P. Roth (1987) for doubts about the empirical adequacy of their studies.

conclusions are hardly forced upon us by the data he cites. Latour assumes that either science is governed by some universal algorithmic scientific method or it is all a social process of negotiation. These are not the only two possibilities. Abstract methods are unlikely to be sufficient to decide scientific debates, and when they have any content, they are unlikely to be universal – this Quinean moral we drew earlier. Yet that does not mean evidence and rationality have no place. Canons of good science, interpreted according to context and combined with background knowledge, can still be a decisive force. Nothing Latour discusses precludes this. In fact much of the "network building" he describes has a natural rationalist reading: the work of others is cited because it is data, challenging other people's network requires doing one's own experiments, i.e., testing, and so on. What Latour describes as network building can be redescribed as a search for evidence and reasons; the network comes in because evidence is holistic. In short, Latour has not established his constructivist conclusions.

So far I have argued that the social constructivists have not made their case. However, their view faces another, even more serious problem. Is constructivism itself a species of scientific activity? Its advocates generally think so. However, isn't all scientific belief caused on its view – and caused in such a way as to exclude appeals to evidence and reasons? The implication then is obvious: we have no compelling reason to accept the constructivist conclusions, no evidence that would require us to believe them. At most social constructivism can influence our beliefs, but not because their beliefs are the most rational or best supported by the evidence. In short, the doctrine appears to be self-refuting. Constructivists are, of course, free to reject the demand for reasons, and then their views will no longer be directly self-refuting. Yet, few social scientists will want to take this step with them, for few researchers really think *their own work* is only a social construction with no claim on evidence and truth as traditionally understood.

I have criticized constructivism on numerous grounds. Yet I should emphasize that my target is its irrationalist conclusions, not the attempt to understand the social processes of science. Constructivist studies bring to light the much-ignored social side of science. Their inquiries can be enormously important for exposing bias and for judging the reliability of scientific practices. This is especially so for constructivist studies of the *social* sciences, where politics and morals so tightly intertwine with the pursuit of fact. Shorn of their irrationalist ideology, constructivist investigations have much to contribute to good science in the social sciences.

I want to turn next to another approach that argues for similar conclusions – namely the "rhetoric of science" movement associated with post-modernism. I shall focus on its most philosophically sophisticated advocate, Richard Rorty.

Starting from Quinean premises, Rorty attacks traditional epistemology. Traditional epistemology holds that "the mind" or "reason" has a nature of its own, that discovery of this nature will give us a "method," and that following this method will enable us to penetrate beneath the appearances and see nature "in its own terms" (1982b, p. 192). Less figuratively, for traditional epistemology we have real knowledge only if we identify the *a priori* constraints on rationality that guarantee that our beliefs accurately mirror the world. Rorty thinks that Quine and fellow travelers Sellars and Davidson have shown that to be a hopeless enterprise. There are no *a priori* constraints and no universal essence of rationality that can serve as the criterion for knowledge. Justification is a holistic, pragmatic affair. Knowledge has no foundations over and above human practice. The idea that our mind "mirrors" reality – that the world has one right description – is incoherent. Truth as correspondence is an appealing but ultimately useless idea. Traditional epistemology is bankrupt.

If traditional epistemology is bankrupt, according to Rorty, then so are traditional conceptions of science. There is no "scientific method" and no special sense in which science is objective, and nothing "called 'scientific status' which is a desirable goal" (1987, p. 42). To be rational is "to simply discuss any topic . . . in a way that eschews dogmatism" (1987, p. 42). The only constraints on science are "conversational"; the only useful sense of objective truth is intersubjective agreement. Thus the only important difference between science and other social practices is one of persuasion versus force. Of course, many human endeavors achieve agreement by persuasion rather than force. So the debate over naturalism rests on a mistake: it falsely presupposes that there is something to be like, an essence of science or rationality.

Rorty's view is seductive. Not only does he raise legitimate criticisms, he appeals to democratic virtues that Western intellectuals admire. Ironically, Rorty's own positive view is both poorly argued for and potentially totalitarian. He mistakenly assumes that the only alternative to the positivist conception of science is an anything-goes sociological approach. Rorty is right that scientific method is not justified *a priori,* that good science is not simply a matter of having the right method, that there may be no single method characteristic of all science, and that the notion of truth as correspondence has proven extremely hard to clarify. However, from these claims it does not follow that science is just persuasion.

As we saw earlier, scientific method can be rationally justified even if it is neither *a priori* nor universal. Rorty assumes that scientific methods either are justified *a priori* or are only conversational constraints. He has ignored a third possibility: that we learn the methods of good science from experience. In fact, we have compelling evidence that mere persuasive power, with no further constraints, is poor reason for belief. Moreover, since most philosophers are persuaded that Rorty is wrong, this is a view he ought to endorse! Of course, this reply assumes skepticism is implausible. If we have no knowledge, then maybe conversational constraints are the only constraints. But then Rorty's claim is only a disguised skepticism, showing his view to be of no independent interest.

Rorty is right that methods do vary across sciences and do not provide a foolproof, mechanical basis for choosing theories. However, as we saw in criticizing Kuhn, changing standards need not be irrational ones. We can have good reason to change our methods: we can learn that they do not promote our cognitive goals. And, as I suggested in Section 2.1, methods will generally not settle scientific disputes *all by themselves*. However, few scientific assumptions will settle disputes in isolation. Embodied in theories and background assumptions, methods can help rationally adjudicate scientific controversy. *A priori* eternal criteria and local rhetorical devices do not exhaust the options.

Rorty's complaints about truth are reasonable. Philosophers have given no adequate substantive account of truth.[18] Assume that Rorty is right that truth as correspondence has no coherent formulation. What follows? We may have trouble showing that science is rational in the sense of getting closer to the truth (defined as correspondence). Nonetheless, as I pointed out in Section 2.2, there are other important rationalist theses. Reasons and evidence can motivate scientific change and theory choice, even if truth is not guaranteed. Likewise, the world, not just conversation, can constrain what we believe even if we have no coherent account of truth as correspondence. In short, empirical or observational evidence can still be the heart of good science. We can show science is rational in this sense without showing that current science is largely true. Rorty mistakenly equates rationalism with its strongest variants.

The social constructivists and Rorty thus point out important inadequacies in past pictures of science. Both err, however, in assuming that the sociological approach is the only alternative. It is not.

2.5 The subtle invasion of values

Lurking behind the irrationalists' views is another, older argument: that there is no fact–value distinction and thus that science is ulti-

---

[18] I assume here that disquotational theories are not "substantive" – they are compatible with correspondence, coherentist, etc. accounts and are less than what strong scientific realism requires.

mately a kind of moral choice. If science does essentially involve moral or political choices, then there are serious doubts about scientific rationality. There are well-known reasons for thinking that moral beliefs are subjective. Even if moral beliefs can be rational, they are not likely to be so in the full-bodied sense that we standardly attribute to natural science. I thus turn in this section to consider challenges to the rationality of science based on fact and value.

Let me clarify some preliminary issues before turning to the main arguments:

(1) The quest for a value-free social science is not a quest for science that presupposes no value judgments. Science essentially involves innumerable judgments about what is good and what ought to be done. However, value assumptions are problematic only if they are moral or political values – as distinct from epistemic values. Reliability, objectivity, fruitfulness, scope, and so on are important values in science, but they are epistemic values. The interesting thesis is thus that science presupposes *non-epistemic* values.
(2) Any interesting argument for value-ladenness needs to show that science *inevitably* makes moral evaluations – that the scientist qua scientist makes moral judgments. Pointing to racial or sexist biases and the like is not enough.[19]
(3) A value-free science does not require that morality is inherently subjective. In other words, the fact–value dichotomy and the objective–subjective dichotomy are not necessarily coextensive. Some or all moral statements might well be objective. So arguments that we can give reasons and evidence for moral claims are not directly to the point.

In sum, those who argue that science is value laden must show that science essentially presupposes non-epistemic values and that these values vitiate the prospects for a neutral, objective scientific enterprise. These may sound like tough requirements. However, they are requirements current commentators claim to meet.

Gunnar Myrdal (1970) argued – long before positivism had exhausted itself – that science did not proceed by taking entirely neutral data and applying a formal logic of confirmation to prove or disprove theories. Myrdal saw that this more complex picture of scientific practice made it

[19] To be fair to the defender of value-ladenness, "unavoidable" here is something short of logical or conceptual necessity. If we can show that values are built in for all practical purposes, then the rationalist position looks defeated.

much harder to sharply separate scientific from moral concerns. Mainstream economic theorizing about underdevelopment, he argued, focused on equilibrium models; by doing so, it emphasized the virtues of free markets and ignored the failings of capitalist development. For Myrdal, this was just one example of a general and unavoidable problem: we develop theories and gather data only after we have decided what are the interesting questions. But, it is not a scientific fact which questions we should pursue. Thus science is inherently value laden. The best we can do is make our value assumptions explicit.

Myrdal's remarks are insightful and suggestive. Nonetheless, he has not shown that values permeate science. Theories do presuppose some prior decision about which questions to pursue. It does not follow, however, that every question we ask is a subjective matter. We pursue some questions because they result from our best theories. "What is the nature of genes and inheritance?" is not just something biologists found interesting, it was a question imposed in clarifying evolutionary processes. Moreover, even when extra-scientific considerations determine which questions we ask, our answers need not be subjective. Once we set the question, it may be a perfectly objective, non-normative matter what the answer is.

A different argument comes from Rudner (1953). Scientific evidence is always a matter of degree. At some point we have to decide if our evidence is enough, whether our predictions sufficiently correspond to the facts to warrant belief, and so on. Nothing in the scientific process, Rudner claims, tells us when the evidence is good enough. Instead, decisions to accept a hypothesis depend on what we would lose by being wrong and gain by being correct. Evidence that is clearly good enough in a circumstance when we have little to lose from being wrong may not be good enough when circumstances are reversed. Thus, accepting hypotheses is inherently value laden.

Rudner's argument turns on an important truth: accepting a hypothesis involves more than its scientific virtue. However, that fact – one emphasized by Kuhn and other post-positivist writers – does not entail that science is value laden. The reason is relatively simple. We can still distinguish scientific or epistemic reasons from normative ones.[20] As real people, we take both into consideration. No doubt when deciding on whether to accept, or act upon, or even entertain a hypothesis, pragmatic or normative consequences are relevant. Yet, their relevance is an independent matter. We can weigh the cost of error, but we must also know the *probability* of error. In short, the degree of evidence and the cost of being wrong are two separate factors, and the former can remain an entirely objective matter.

---

[20] This line of argument is advanced by McMullin (1982).

The arguments of Myrdal and Rudner ultimately turn on the holistic nature of belief. Longino (1990) has recently argued directly from the holistic nature of testing and theory to the conclusion that values are a necessary part of science. Following Quine and many others, Longino argues that data by themselves do not tell us what to believe, for their implications and even description depend on background assumptions. So no scientific method can guarantee that values are removed from the scientific enterprise. Values are ultimately built into science, because they are part and parcel of the assumptions we use to make science. Does this then make science merely a reflection of individual values? Longino does not think so. Science is essentially a social enterprise – scientific results are subject to public scrutiny and check. This assures us of an intersubjective constraint that removes subjective *individual* preference from science.

Longino nicely identifies ways in which values can infiltrate science. She also argues convincingly that some research on gender differences reflects various androcentric or sexist assumptions. She correctly points out that, given the holism of testing, value-free hypotheses tested against value-free data may still involve value-laden background assumptions. However, this does not show that *good* science can be value-laden science. Good science requires more than testing factual hypotheses against neutral data. It also requires that the theory of the test meet scientific standards as well. When statistical inference, for example, rests on unsubstantiated and biased assumptions about the relevant causal variables, linearity, and so on, bad science is at work. However, the formal rules of statistical inference do not prevent such sloppy research. That does not mean we must count such research as good science, only that we must evaluate it on substantive and empirical grounds. Background assumptions supported only by moral judgments make for bad science. However, if we look at the research Longino cites, biased or bad science is what we see: research based on shaky, often unacknowledged background assumptions that reflect androcentric stereotypes. Nothing here shows that values can or must permeate good science.[21]

So none of these arguments show that non-epistemic values inevitably influence science. Of course, if natural science is not inherently value laden, *social* science might be. We will consider some arguments for this idea in later chapters. However, even if values do not inevitably penetrate

---

[21] Longino (1990, p. 100) has another argument based on the idea that values are built into "the constitution of the object of study." The idea seems to be that in defining a scientific domain non-epistemic values are inevitable. This raises issues too complex to deal with here, but I am unconvinced by this reasoning as well – see Kincaid (in press b), where I discuss such arguments in the context of economics.

social science, nonetheless *bias* is a much more troublesome problem in the social sciences than in the natural sciences. Longino and the others have helped me see just how subtly values can penetrate science.

2.6 **The symptoms of good science**

This chapter set out to do three things: answer irrationalist critics, sketch some post-Kuhnian philosophy of science, and identify features of good science. We have addressed directly the first two tasks. I want now in this final section to take up the third task. Doing so is crucial to defending the naturalist position advocated throughout the rest of the book.

To show that the social sciences can be scientific, we need some account of what being scientific comes to. Time and again I shall argue in later chapters that the social sciences can achieve the standards of good science and that some research actually has done so. However, attempts to identify the defining features of science have a long and disappointing history. Furthermore, the view of science and scientific method I used in answering the irrationalists seems prima facie incompatible with defining good science. If there is no purely formal, *a priori* scientific method but only contingent, historically specific empirical evaluations that must be embodied in concrete theories, how can we say anything useful about "the" methods of good science? How can we say that areas as diverse as molecular biology and sociology use the same methods or have the same virtues? Thus before discussing the traits of good science, I need to address these procedural questions.

Past attempts to define science – such as Popper's falsifiability criterion – have indeed failed. The positivists' goal, however, was much more ambitious than anything needed for the argument advanced here. I make no claim to have the individually necessary and jointly sufficient criteria that uniquely divide all inquiries into two groups, namely, the scientifically respectable and the pseudoscientific. Instead, the criteria I advance are the symptoms of good science; they are indicators of good science that make no claim to completeness or perfection. To argue for naturalism, I need not show that there is a sharp distinction between the scientific and pseudoscientific. Rather, I need to show that the social sciences share the basic virtues common to paradigm instances of good science – that the social sciences are on the scientific end of the science/non-science continuum, if there be such.

The naturalism I defend here also does not argue that the social sciences meet criteria that characterize good science throughout history. There might be no such criteria, even loosely construed. My argument seeks "only" to establish that the social sciences can be sciences according to criteria we *now* take to define good science. (The rationalism I am presupposing is a moderate one – recall the discussion of Section 2.2.) That conclusion, however, is still highly contentious. And if current criteria for good science do reflect universal standards, my arguments entail that the social sciences can meet them.

Furthermore, areas as different as molecular biology and sociology can share the same virtues. It is useful here to distinguish between *abstract* and *realized* virtues. Scope, simplicity, accuracy, falsifiability, and so on are abstractions in that they do not directly mention specific empirical presuppositions. Realized virtues, on the other hand, are the ways a specific science embodies abstract virtues in the relevant empirical detail. Thus simplicity in nineteenth century geology became a principle of uniformitarianism;[22] falsifiability or objectivity in current clinical research is realized in double-blind experiments, for example.

Obviously, at the most concrete level, the social sciences do not and cannot share the virtues of the natural sciences. Yet, at the most concrete level even the subdisciplines of physics do not share the same methods either. Nonetheless, it can still be the case that (1) there are broad virtues at some level of abstraction that characterize both the natural and social sciences and (2) the social sciences can and do successfully embody or realize those virtues. Both propositions are essential. Abstracting broad virtues from their concrete applications allows us to identify the core of scientific rationality and to avoid concluding that there are as many scientific methods and virtues as there are scientists. Yet simply identifying abstract virtues like scope and fruitfulness is not enough. A convincing argument requires showing how those virtues get embodied in social science research, a lesson we also learned from Kuhn. This two-sided approach avoids a universal *a priori* logic of science without denying that there are general standards of scientific rationality.[23]

Nonetheless, it might turn out that no unitary set of virtues emerges even after abstracting from contextual details. As we will see shortly, that may well be the case for non-evidential constraints on scientific rationality. Yet we can still defend naturalism. Naturalists can grant that there are multiple ways to do good science. However, they cannot allow that those methods are all beyond the reach of the social sciences. Nor can they allow that the social sciences have special methods that do not embody some basic scientific virtue guiding the natural sciences. One common substantive set of scientific virtues makes arguing for naturalism easier, but it is no prerequisite.

What then are the symptoms of good science? For our purposes it will

---

[22] I take this point from Sober (1988).
[23] Here I thus disagree with Rouse (1987, p. 124).

be useful to break "science" into two components – the process of scientific inquiry and the products of that investigation. Traditional philosophy of science generally concentrated on the latter, namely, on theories and their relation to the evidence. However, Kuhn and the social constructivists have highlighted just how one-sided that emphasis is. Science is above all an activity, and evaluations of good science that ignore that fact are at best incomplete. Thus the symptoms of good science involve both the symptoms of good products and the symptoms of good processes.

Traditionally, philosophers of science have identified symptoms of a good scientific *product* that fall into three broad categories: (1) evidential, (2) explanatory, and (3) formal. Evidential virtues concern how well data support a theory. It is these virtues that philosophers of science emphasize. However, science aims for more than true or well-confirmed statements. A phone book contains numerous truths, but it lacks an essential scientific virtue: the ability to explain. Good science does not just describe, it explains.[24] Finally, it is sometimes thought that good science does more than just provide true or well-confirmed explanations – it produces *theories* with certain formal properties. Theories, on this view, are systems, where a "system" is usually equated with explicitly stated bodies of propositions, universal in scope, that can be axiomatized and deductively organized.

The symptoms of good scientific *practice* are those that indicate reliability – in short, they show that a given practice brings about good scientific products. Philosophers of science have ignored this side of science, so we have no standard typology. We can categorize good processes by the different components of reliability: a process is reliable if it produces good products in large quantities or efficiently or quickly and so on.[25] We can also classify virtues depending on whether we describe them epistemically or non-epistemically. An "impartial" process, for example, is one that is epistemically described, for it is tied to producing a good scientific product. A "competitive, interest-driven" process has no such connotation, though *a priori* it is not precluded from reliably producing good science as well.

Throughout this book I shall focus primarily on three kinds of symptoms: evidential and explanatory virtues of scientific products and the epistemically described virtues of social processes. I downplay other kinds of virtues for two reasons. Chapters 3 and 7 will argue that formal virtues

[24] Of course some philosophers of science – instrumentalists – may argue that good science is only empirically adequate, yet they too typically recognize that explanation is in some sense essential to good science and thus attempt to show how explanatory virtues can be collapsed into evidential ones.

[25] Here I am borrowing from Goldman (1986).

are relatively unimportant; good science does not require deductively systematized theories of universal scope. Thus I will not discuss those virtues any further here. I downplay the symptoms of good processes for a very different reason: we have no well-developed account of what processes produce good science, at least when those processes are not described epistemically. Psychology and sociology have much to learn about what factors make for good science, and I thus cannot draw on an established body of knowledge. That does not mean we can say nothing. Chapter 8 will make some limited claims about these topics when we discuss why the social sciences are not more effective. However, those judgments are tentative. Thus the main argument for naturalism will concentrate on showing that the social sciences can and sometimes do use good methods to produce well-confirmed explanatory accounts.

With these methodological preliminaries out of the way, let's turn now to look concretely at the symptoms of good evidence and good explanation. We can describe these virtues at different levels of abstraction. At the most abstract level are the standard, one-word virtues such as scope and fecundity. At the most concrete level, these symptoms become double-blind experiments, Durbin-Watson tests for autocorrelation of time series data, and other such domain-specific practices. Between these two extremes are generalizations about good science that hold across disciplines and yet are more informative than the simple list of scientific virtues. What, for example, are the general practices that specify empirical adequacy, fruitfulness and objectivity? In what follows I discuss the idea of good science at the first two levels; the third level will have to wait for later chapters when specific empirical work is at issue.

At the most abstract level, good science typically has at least the following *evidential* virtues:

> *Falsifiability.* While falsifiability is far from the entirety of good science, it is the first line of empirical adequacy.
> *Empirical accuracy.* The more predictive success, the better – measured in terms of both quantity and quality.
> *Scope.* We want theories that not only repeatedly predict the same kind of phenomena with precision but also that predict a wide variety of different kinds of phenomena.
> *Coherence.* A good theory coheres with our best information from other sciences. Logical consistency is a first start, tight evidential interconnections, the ideal.
> *Fruitfulness.* Theory evaluation is not just evaluation at a time; it requires looking at past track record and future promise.

Objectivity. Standard lists of scientific virtues seldom explicitly mention objectivity, apparently assuming that it falls out from the above traits. But Kuhnian worries about incommensurability and doubts about value-free science should suggest that empirical success may not guarantee objectivity. Nonetheless, the ideal of an unbiased, disinterested pursuit of the truth is the hallmark of science. Science is objective when our beliefs reliably indicate the way the world is rather than the way we want the world to be.[26]

These traits help separate good science from bad. Yet, they are also abstract and simplistic. By themselves, these standard criteria will seldom resolve disputes over scientific adequacy – because they admit of multiple interpretations, because we do not know how to measure them or trade one for another, because they are simplifications that hold only *ceteris paribus*. Hence we need to move to the more concrete practices that realize these virtues. In particular, I want to focus on three broad features of good empirical tests. Good science undergoes *fair tests, independent tests,* and *cross tests.* Before we spell out these tests in detail, let me explain why these traits are important.

On the most philosophical level, being tested against experience is the basic defensible idea from empiricism. If empiricist accounts of meaning, of demarcation, and of theory and its relation to observation ran into fatal criticism, the idea that knowledge requires testing beliefs against experience did not. Fair tests, cross tests, and independent tests are a concrete embodiment of that empiricist tenet. Testing virtues are also important because they bring in the active side of science. Focusing on the relation between observation statements and theoretical propositions, as so much philosophical confirmation theory has, violently abstracts from real science. Appeals to "the data" summarize a history of hypothesizing and manipulating the experimental conditions that lead to the end product. Focusing on testing virtues acknowledges that real confirmation is essentially a process and thus gives us virtues of scientific practices as well as scientific products.

Of course, the best rationale for focusing on these testing virtues would be a full empirical theory of methodology replete with detailed evidence. I clearly have no such theory nor is there one extant in the literature to

---

[26] This intuitive idea, which lies behind the more concrete traits that I discuss below, has its origin in psychologistic approaches in epistemology and is developed in connection with philosophy of science by Miller (1987).

draw upon. Nonetheless, the kinds of tests I describe here do loom large in several careful recent accounts of modern experimental physics, namely, those of Franklin (1986, 1990) and Galison (1987). Their work also makes some effort to show that appeals to such tests are not merely rhetorical devices in a social struggle between scientists. I can offer no further support here, though presumably readers will find these tests characteristic of the natural science they know best.

Let's look now at these testing virtues, starting with *independent* tests. An independent test is one which tests the hypothesis at issue rather than the entire theory to which it belongs. As we saw in Section 2.3, the holistic nature of theories does not preclude directing tests against specific hypotheses. To do so the theory of the test and the hypothesis at issue should be logically independent. The theory of the test includes all the background assumptions needed to generate data relevant to the hypothesis at issue. Those assumptions and the hypothesis being tested are logically independent when the falsity of either one would not entail the falsity of the other. In short, in an independent test the hypothesis does not presuppose the truth of the theory used in constructing the instruments, identifying the initial conditions, interpreting the data, and so on, nor do those theories presuppose the truth of the hypothesis.

Tests are independent in a second, equally important sense when the theory of the test has its own evidence. If we accept the theory of the test for the same reasons we accept the hypothesis at issue, then our test is compromised – especially if the test assumptions are improbable given what else we know. In other words, in the ideal test the hypothesis is both logically and evidentially independent of background theory. The physics that grounds electron microscopy does not depend on what we know, for example, about how cells ingest large proteins.[27] So when an electron micrograph shows a cell wall enclosing a particle, we have a quite independent test of cell biology. If, on the other hand, our only reason for believing in the data from EMs was the same as that for the endocytosis hypothesis, our test would not be independent.

Independent tests thus give content to the abstract virtues listed above. Theories without independent tests are hard to falsify. They also obviously lack objectivity, for the evidence for the hypothesis depends essentially on our background beliefs, not the world. Lack of independent tests also

---

[27] The situation is, however, more complex than this statement indicates. If I have independent evidence for endocytosis in a given cell line, then successful EM depiction of that process is evidence in turn for the reliability of EMs. Obviously, a more sophisticated account calls for identifying degrees of independence, primary evidence, and the like.

indicates lack of coherence, for the theory being tested is only integrated with itself.

Good theories also survive *fair tests*. The basic idea behind a fair test is that a good theory rules out its competitors in an objective fashion. Fair tests must first be independent tests. Theories that presuppose their own truth are not winning fairly. But independent tests are not enough. Fair tests depend crucially on competitors sharing sufficient background to pick a winner. That requires sharing, first, language, and second, the assumptions essential to the theory of the test. Section 2.3 discussed how different theories can share language. Ideally, both theories share a range of terms used in constructing the test. When that is not possible, theories must at least be translatable – there must be some way, perhaps complex, to capture the reference of key terms. Beyond sharing language, a fair test requires that no assumptions are made that bias the test towards one hypothesis rather than another. Competing hypotheses need not share all background theory, but they do need to share enough to generate unbiased tests.

Thus theory evaluation is essentially comparative. Even if we had some metric for virtues like predictive accuracy and scope, those measures would not be good enough, because they do not ensure a fair test. Without common data and assumptions, simply counting up successful predictions would not tell us which theory was preferable. Of course, no individual test is by itself logically decisive. But, as Lakatos (1970) saw, repeated failures of fair tests tip the scales. Fair tests are thus the essence of scientific objectivity.

Finally, good science is *cross tested*. Cross testing involves numerous procedures – for example, triangulation, bootstrapping, and data-reliability analysis – that are part and parcel of good science. The root idea is roughly to make sure that it is the world, not our background assumptions, which leads us to accept a hypothesis. Just as we use Mill's methods of variation to isolate causes, so too varying the different components of a test helps show that our hypotheses have really been tested. Cross testing therefore involves varying some assumptions while holding others constant. Ideally we would like to test a particular hypothesis using diverse background assumptions. Similarly, we would like to see the same theory of the test employed to investigate a variety of different hypotheses. For similar reasons, multiple measures of the data are desirable. These processes put the holistic nature of theory and evidence to good use. Far from entailing that theory evaluation is subjective, the web of belief places numerous constraints on theory acceptance. As evidence, hypotheses, and background assumptions are varied and checked against each other, the odds diminish that successful tests happen just by chance.

The modern theory of the gene nicely illustrates cross testing. Postulated to explain the facts of inheritance, the gene hypothesis is cross-checked by embedding it in what we know about cell functioning, biochemistry, and evolutionary change. Biochemistry of course makes innumerable predictions other than those about genes and is cross-tested against our background knowledge of both physics and cell functioning. Cell biology is in turn tested by what we know about chemistry and physics (as in the electron micrograph example), physiology, anatomy and the like. Finally, the gene postulate likewise generates specific predictions when combined with evolutionary theory. Of course, these areas are really large complexes of diverse claims and evidence, independent in varying degrees. As a result, the actual cross checking is enormous.

Like independent and fair tests, cross checking also gives substance to the more abstract goals of empirical accuracy, scope, and coherence. Replication and randomization are varieties of cross checking that try to vary *unknown* background assumptions. Diverse kinds of evidence are important, because different evidence requires varying the theory of the test. Integrated theories are important not simply or even primarily because we value organized wholes; more important, integration forces us to cross-check, upping the odds of truth. And, while not enough to ensure objectivity, cross checking certainly makes bias more difficult. When combined with independent and fair tests, cross checking is a powerful method.

These three kinds of tests give us a more concrete guide to good science. They will play a key role as we defend naturalism in later chapters. There I shall argue both that the social science research can provide such tests and that no current social research confirms without doing so. There can be and is good social science research – and it is the result of fair, independent, and cross testing.[28]

I want to turn now to symptoms of good scientific explanations. Numerous such traits are cited by philosophers and scientists. Identifying mechanisms, providing relevant laws, citing unobservable entities, unifying diverse phenomena, and describing causal processes are mentioned most frequently. Unfortunately, all of the above constraints have been dis-

---

[28] These three virtues have a natural Bayesian reading. Fair tests ensure that alternative explanations are used to calculate the probability of the data, among other things. Cross tests provide ways of determining whether $p(E/H)$ really rests on auxiliary assumptions, of assigning a low probability to the hypothesis that the data result from artifact, of raising $p(H)$ by showing it coheres with other work, and so on. Independent tests are ones where $p(H/A) = p(H)$, and A is the auxiliary test assumptions. It can be shown that with such an assumption the evidence will bear differently on the hypothesis and the theory of the test. See Howson and Urbach (1989, pp. 96–102).

puted at one time or another — basically because philosophers disagree about what explanation itself is. Thus it is harder for us to give any noncontroversial account of what promotes good explanation.

Most approaches to explanation do, however, fall into two broad camps: those that emphasize causation and those that emphasize unification.[29] On the latter view, explanation comes from showing how diverse phenomena fit into a common pattern. When we can account for phenomena by invoking fewer assumptions, we improve our ability to explain. Newton took the apparently diverse phenomena of tides, projectiles, and planetary motion and showed how they all followed from one set of principles. Darwin did the same for the many different traits and behaviors of plants and animals. Explanation seems to come from unification.

If explanation as unity emphasizes the global, the causal approach starts with more local facts. Science at its best gives us well-confirmed accounts of particular causal processes. It is those processes which the experimental method best illuminates and which tell us how the world works. Theories of broad scope may be possible, but their explanatory power comes from citing causes. Particular causal processes may be instances of general causal laws. Low-level causal generalizations may in turn hold because of more fundamental and general causal processes. So the causal approach can allow unifying laws, both fundamental and derived. Yet those laws are explanatory only because they summarize the causal facts.

Both approaches, I should note, can allow that pragmatic factors play a role in explanation. Philosophers have had some success in clarifying pragmatic factors in explanation (Achinstein 1980; Garfinkel 1981; van Fraassen 1981). One helpful suggestion is that explanations are answers to questions. Questions, however, do not carry their meaning on their face: the same question can call for different answers depending on the context. One contextual factor is the *contrast class:* "Why did Adam *eat* (rather than throw, etc.) the apple?" calls for a different answer than "Why did *Adam* (rather than Eve, the snake, etc.) eat the apple?" Philosophers have developed these ideas in some detail.

The causal and unification views incorporate these insights as helpful emendations. For the causal view, different questions and different background knowledge means focusing on different strands in the causal net.

---

[29] This division is not the only way to categorize different accounts; Wesley Salmon's (1984) epistemic vs. ontic classification, for example, is similar to but not the same as the unification–causation dichotomy, for there might be, for example, other real-world relations besides causation that explained. Nonetheless, my discussion of explanation here and elsewhere in the book draws heavily on Salmon's general approach.

When the highway engineer and the car manufacturer ask what explains the accident rate, they may want different answers. Those answers focus on different parts of a complex total cause, but they are still nonetheless *causal* explanations. The unification view can take a similar tack. Pragmatic factors may determine the connotation of questions and the relevant kind of answer. But an answer that explains is still one that provides the most unified story.

How do we go about defending naturalism, given these two different pictures of explanation? I can see three basic approaches: (1) The "separate but equal" strategy opts for a plurality of explanatory virtues. Causation is the key to explanation in some domains, unification the key in others. If they go this route, naturalists must then argue that the social sciences can and must provide good explanation by one of these two standards. (2) The "unity in diversity" strategy would argue that causation and unification are really two compatible aspects of good explanations.[30] Causation emphasizes that part of explanation that involves an objective relation in the world, what Salmon calls the ontic notion of explanation; unification focuses on the epistemic side of explanation – it clarifies the idea that explanations make something more familiar, expected, or understandable. Naturalists taking this route then would ideally like to show that the social sciences can and must produce explanations that are causal and unifying. (3) The "eliminativist" strategy would argue that one or the other virtues was really derivative – that causation only explained when it provided unity or vice versa. A naturalist who adopted this strategy would thus argue that the social sciences could and must have good explanations by whatever criterion was basic.

My defense of naturalism will combine these three strategies. My main concern will be to show that the social sciences can produce well-confirmed causal accounts. I will forestall worries that unification is the essence of explanation in several ways. First, I shall assume that citing causes does explain and thus that any theory of explanation must be consistent with this fact. Since defenders of unification typically grant this point, my assumption is a safe one. Secondly, I shall argue in the next chapter that unification is not *necessary* to explanation and probably not sufficient. So showing that the social sciences can give causal accounts is showing that they can explain.

What about the second naturalist thesis – namely, that the social sciences *must* proceed along natural science lines? I will adopt the "separate but equal" strategy and argue that all social science explanations involve

---

[30] Salmon (1989, p. 180) has suggested that some such reconciliation might be possible.

either unification or causation. The social sciences, in short, have no special basic routes to explanation. When social theories unify but do not cite causes, our assessment of them is tentative. If inquiry ultimately shows that explanation is unification, those theories will explain. If, as I would suggest, citing causes is the prototype for explanation, then those theories may serve other roles. However, whatever the outcome, the same conclusion will hold for the natural sciences, since unification of course plays an important role there. So explanation will draw no sharp divide between the natural and social sciences.

Lest it seem that I will be arguing for a radical monism, let me emphasize a point made in Chapter 1: basic scientific virtues common to all science are compatible with great diversity in scientific practice. The virtues described here are abstractions. Those abstractions may be embodied in many different specific explanatory strategies and confirmational methods. In fact, our Quinean philosophy of science expects such a plenitude. For example, even if the social sciences always gave us causal explanations, it could nonetheless gives us many different *kinds* of causal accounts. Explanations emphasizing structural causes, causal redundancy, microcausal details, functional causes, and so on might well be appropriate, depending on our background knowledge. We can allow and indeed expect a similar diversity when fair tests, cross tests, and independent tests are embodied in real research. Thus nothing in my account rules out "methodological pluralism," properly understood.[31] Naturalists deny only that those methods are outside the realm of natural science – that they are in no significant sense embodiments of basic standards of good science in the natural sciences. So arguing for naturalism requires seeing exactly how and if social scientific practice realizes these abstract virtues. It is to that task which we turn next.

[31] As, for example, defended by P. Roth (1987).

3

Causes, confirmation, and explanation

This chapter begins the argument for a *science* of society. I argue that the social sciences can produce well-confirmed causal explanations and laws. No inherent conceptual obstacles prevent the social sciences from providing good causal explanations. Nor are there any insurmountable practical obstacles either. Complications abound, of course, but some good social research manages to overcome those difficulties as well as good work in the natural sciences.

My major concern shall be to defend causal explanations rather than laws. Although I think laws play a secondary role in good science, it is important for my overall argument to defend laws as well. Not all philosophers share my skepticism about unificationist accounts of explanation; they will think laws important because they unify. Furthermore, though the exact role of laws in confirmation is also controversial,[1] prima facie laws seem to play an important role in testing. Finally, I shall argue later that in practice there is no very important difference between laws on the one hand and generalizations on the other. So defending causal explanations and defending causal laws naturally shade into each other.

The chapter is organized as follows: Section 3.1 takes on conceptual arguments coming from Searle, Davidson, Taylor and others. After disposing of those objections, Section 3.2 examines a much more serious and practical obstacle to causal laws – the fact that most alleged social laws are qualified *ceteris paribus*. Generalizations with open-ended escape clauses look unfalsifiable and unexplanatory. Section 3.2 argues that these qualifications are ubiquitous to all science and outlines the various ways good natural science deals with such problems. I then look at some recent

---

[1] For example, Cartwright (1989) argues that singular causes, not laws, are fundamental. That view, however, is controversial. See Papineau (1991).

social research – namely, Paige's work on agrarian political movements – and argue that his research handles the *ceteris paribus* problem as well as some of our best work in biology. Section 3.3 turns to more general doubts about non-experimental evidence, which is rife in the social sciences. Finally, Section 3.4 takes up worries that the social science I defend does not explain because it lacks unifying laws and systematic theories.

3.1 Some *a priori* objections

Philosophers have given numerous reasons for thinking social laws impossible. Some point to the alleged fact that the basic kinds or predicates of the social sciences have no determinate physical definition. Others think that the social sciences will never produce real laws because they cannot describe all the causes operative in their domain. Still others argue that humans are "self-defining" and that this fact precludes causal explanations. In this section I sketch these arguments and argue that all fail.

John Searle (1984) has recently argued that social laws are impossible because social kinds or categories have multiple realizations. According to Searle, "the defining principle of . . . social phenomena set no physical limits on what can count as the physical realization" (p. 78). "Money," for example, is through and through social in nature – its definition refers to its social function, not its physical attributes. As a result, nearly anything can serve as money. Most or all social kinds are in the same boat: they have indefinitely many diverse physical realizations. This "means that there can't be any systematic connections between the physical and the social . . . therefore there can't be any matching . . . of the sort that would be necessary to make strict laws of the social sciences possible" (p. 79). Other philosophers (Churchland 1979; Rosenberg 1980, p. 107) have offered similar arguments to show that "folk" psychology – explanation in terms of beliefs and desires – is bereft of laws.

Let's tighten up Searle's argument a bit. He seems to be saying something like this:

(1) Social kinds have indefinitely many physical realizations.
(2) When a kind has indefinitely many physical realizations, it has no systematic connection to the physical.
(3) If a kind is not systematically connected to the physical, it cannot support genuine laws.
(4) Thus social kinds cannot support genuine laws.

This argument is, so far as I can see, either invalid or unsound, depending on how we read "systematic connection." If a systematic connection requires a law-like relation between social and physical predicates, then premise (3) makes a highly implausible claim: that the social sciences cannot produce laws unless they are reducible to physics. We shall discuss and reject such reductionist programs in Chapter 5.

However, we do not need that discussion to see that Searle's requirement is ill conceived. If reduction to physics is required for laws, then large parts of the natural sciences cannot produce laws either. The fundamental predicate in population genetics and evolutionary theory, namely fitness, clearly has no unique physical definition, for an organism can be fit in indefinitely many physical ways (Rosenberg 1978). So there is no single or even complex physical characteristic which we can use to define "fit." The same problem surfaces in molecular biology for terms such as "antibody," "signals," or "receptors," which all have open-ended physical realizations.[2] Even "planetary body" in Kepler's laws seemingly has no unique definition in quantum mechanical terms. So if we read "systematic connection" as the type-type connection of reduction, much natural science cannot produce laws. Searle's argument shows too much.

Perhaps we should read "systematic connection" as a looser relation, namely, supervenience. One set of facts A "supervenes" on a set B roughly if fixing the B facts also fixes the A facts. Premise (3) becomes more plausible on this construal, since *reduction* to physics is no longer required for laws (reduction requires more than supervenience – see Chapter 5). However, premise (2) is now false, for supervenience does not rule out multiple realizations. For example, physically identical organisms will have the same fitness, but there are indefinitely many physical ways to be fit. Thus either premise (2) or premise (3) is false, depending on how we understand "systematic connection." Equivocating on the term would, of course, render the argument invalid.

Another common argument (Davidson 1970; Taylor 1971) turns on the fact that the social realm is not "closed." Laws by nature must be universal. But if the social world constitutes an open realm subject to outside forces – physical or biological events for example – then social theory will remain forever incomplete and forever without true laws. Laws seem precluded.

We can take this argument in two ways: as arguing (1) from a fact about social systems as objects in the world or (2) from an alleged fact about social theories, namely, that they do not cover all the forces or causes in their domain. Neither rendering produces a sound argument.

Social systems – a given institution, society, or even world system – are obviously not closed systems. They depend on both physical and biologi-

---

[2] See Kincaid (1990d).

cal factors. However, every physical system short of the entire universe is also influenced by outside causes. So merely describing open systems cannot preclude laws. If it does, then the only laws in physics are those that describe the totality of the universe, another unacceptable conclusion, I assume.

Perhaps the crucial issue concerns not the open or closed nature of actual systems but instead a theory's ability to handle those outside factors. A closed theory is complete: it can describe and explain *in its own terms* all the forces acting in its domain. So, the argument runs, forces affecting open physical systems can be fully handled within physics itself. In the social sciences, however, outside factors are not social in nature and thus social theory cannot handle them. Consequently, alleged social laws are bound to be incomplete.

This argument again threatens to prove too much, namely, that no physical laws are possible either. Biological, psychological, and social events influence the physical universe, thus creating apparent exceptions to physical laws. If, however, our biological, psychological and social theories are even in part irreducible to physics, then we cannot handle these exceptions in entirely physical terms. Physics will be incomplete as well – and thus, if this argument is right, without laws. I again assume that conclusion is unacceptable.

The open nature of social science carries little weight for another reason. Why does a real or strict law have to invoke language only from one theory? The above argument assumes it must, but that assumption seems quite unwarranted. Cellular biology invokes chemical facts, and evolutionary theory does the same with physical facts about the environment. Why should that undercut laws if the law identifies kinds and relates them in whatever manner laws require? This problem is particularly acute because each physical law invokes only a subset of the total physical language. When other *physical* forces interfere, that subset obviously will not have the vocabulary to handle this more complicated situation. Are those laws only apparent because they are not refinable in their own terms? Here it just seems silly to make lawfulness turn on some prior notion of the "right" vocabulary. The real issue is whether potential exceptions can be handled in a systematic way. Thus there might well be social laws even if they are not refinable in a purely social vocabulary.

A final influential argument turns on the idea that humans are "self-defining." Taylor (1971), Fay (1984), and numerous others have advanced versions of this argument. The basic idea is that how we categorize our behavior determines in part what that behavior is. As autonomous agents, humans change their self-conception. As a result, any causal generalizations about the social world are bound to be flawed and without real nomic force. Only by completely ignoring social context and treating human behavior as brute motion could we ever develop such laws. So the reasoning goes.

We can take this argument in several different ways. In particular, these doubts might be about applying natural science *methods* to human behavior or about *laws* in the social sciences. I will take up the former kinds of doubts in Chapter 6 when I discuss interpretation. Here I want to focus on this reasoning as a more direct objection to laws.

Two different arguments seem to be lurking here. One builds on the contextual nature of human behavior. Bare bodily motion is not the object of the social sciences. Rather, it is behavior in a social context. Since social context is intertwined with our self-conceptions, and since self-conceptions vary enormously through history and across cultures, then we should expect no social laws. Laws are universal, but human social behavior is contextual.[3]

Once spelled out, this reasoning is uncompelling. Assume human behavior is contextual. That prevents laws only if no science produces laws. When gases become fluids, the gas laws no longer apply and we need different laws for the new situation. thus the gas laws are specific to context. So are innumerable other generalizations in the natural sciences. If appeal to context prevents laws, then the natural sciences have few or no laws either. Thus reference to contexts does not rule out laws.[4]

A second version of this argument appeals to free will. One problem with the previous argument is that even if humans define themselves, those definitions might change in predictable ways. So appeal to context alone must be inconclusive. This suggests that critics are really focusing on the idea that humans are free. Certainly Popper's (1982) attack on laws of history had some such motivation. So this revised argument says that human behavior depends on self-definition and self-definition is a free human product. Thus human behavior cannot be law governed.

There are obvious replies to such an argument, replies so familiar in the philosophical tradition that I shall just mention them: (1) at least some conceptions of freedom, for example freedom as lack of compulsion, are entirely compatible with laws of human behavior; (2) the idea that humans are free in some other sense is notoriously hard to clarify, since uncaused events sound like random events which fit no one's notion of freedom; and (3) even if humans are free in the sense of uncaused, it might still be the

---

[3] This argument is clearly at work. Giddens (1979, ch. 7).

[4] Of course contextual factors can limit the scope of any alleged law. I take up these separate worries later in the chapter and in Chapter 6. For a helpful discussion of these issues see also Little (1989, chs. 6 and 7).

case that large-scale human behavior is patterned, just as indeterminacy might hold at the subatomic level and Newton's laws at the level of medium-sized objects. Thus this version of Taylor's argument also fails.

No doubt there are numerous other arguments alleging that social laws are impossible. Some of those arguments will come up in later chapters on interpretation and functionalism. At this point I want to turn to a much more serious obstacle to social laws – namely, the fact that most alleged social laws are qualified *ceteris paribus*.

3.2 Confirmation and qualifications

Eliminating conceptual objections is, of course, only a first step. While a science of society might be possible, it might be only that. In reality social research might never achieve even minimum standards of scientific adequacy. Having answered *a priori* objections, I now begin a more fundamental task: showing that the social sciences *in practice* can and sometimes do produce well-confirmed causal explanations. This section begins that argument by focusing on one serious practical obstacle. Almost every candidate for social laws is qualified *ceteris paribus* – "other things being equal." For critics, however, *ceteris paribus* clauses look like open-ended escape clauses, making the social sciences seem unfalsifiable and non-explanatory. That appearance is deceptive, or so I shall argue. In the process we shall make a prima facie case that some research actually does produce well-confirmed causal explanations.

3.2.1 *How can* ceteris paribus *laws be confirmed and how can they explain?*

Nearly every purported causal explanation in the social sciences is implicitly qualified with "other things being equal" or "assuming nothing else interferes." These escape clauses raise doubts about whether social generalizations can be confirmed or can explain. The problems about confirmation are obvious. We may never observe a case where other things really are equal. The weather, for example, influences the economy by numerous routes and the economy in turn influences every aspect of society. Nonetheless, almost no social research makes weather a variable; it falls under the implicit *ceteris paribus* clauses. If, however, we never observe social factors without such disturbing influences, how can we have compelling evidence for any social explanation? It looks like no social generalization could ever be disproven. After all, disconfirming evidence might show only that other things were not equal. And when we do find evidence for a social law, we apparently do not know how to distribute credit. Our law might be predictively accurate only because other things were not equal. *Ceteris paribus* generalizations seem unconfirmable.

There are also puzzles about how *ceteris paribus* laws *explain*. Since in the real world other things generally are not equal, explanations qualified seem irrelevant. They describe how things would behave, not how they do. But, we want to explain this world, not some other possible world. More concretely, if there are interfering factors, then how do we know that the causes our law picks out are the operative ones rather than the interfering factors? How do we tell the genuinely explanatory laws from the irrelevant ones – since strictly speaking no *ceteris paribus* law literally applies? Since *ceteris paribus* qualifications are rife in the social sciences, these questions are pressing.

Ceteris paribus qualifications surely do plague the social sciences. That, however, does not separate them from the natural sciences, for *ceteris paribus* clauses are endemic even in our best physics.[5] Fundamental physical laws describe single forces in isolation. When we explain, however, we are usually faced with a complex physical process where several forces are at work. Only rarely can physics completely flesh out the needed *ceteris paribus* clauses in a systematic, theoretic way. Instead, physicists employ numerous ad hoc principles and rules of thumb to tie *ceteris paribus* clauses down to reality.

A similar point can be made for other "hard" sciences, for example, molecular biology. Molecular biology produces numerous causal generalizations about cells – among them, that cellular processes are produced by the DNA-to-mRNA-to-protein process and that internal cell functioning is stimulated by external tissue through signals bound at the membrane that are amplified by a second messenger carrying the signal to its destination. Yet, these generalizations are really simplifications of more complex processes. Explaining actual cellular events requires factoring in cell type, the kind of organelles involved, the source of the external signal, and so on. Most causal generalizations in molecular biology are qualified *ceteris paribus*.

Ceteris paribus clauses are therefore no *inherent* obstacle to well-confirmed causal explanations in the social sciences. However, defending science in the social sciences still requires answering three questions about *ceteris paribus* laws: (1) How can they explain at all? (2) How does good work in the natural sciences deal with the *ceteris paribus* problem? and (3) Can the social sciences employ similar strategies? I begin with the first, more philosophical question and then turn to look at these problems in practice.

---

[5] For the role of *ceteris paribus* clauses in physics, see Cartwright (1984) and Hempel (1988); see Kincaid (1990b) and M. Salmon (1990) for an application of this point to the social sciences.

Ceteris paribus generalizations cause problems because things are seldom equal, yet we use those laws to explain. If other things are not equal, then our explanation is strictly speaking false. So how can a *ceteris paribus* law explain? We cannot answer this fundamental question without some picture of explanation. In the last chapter I sketched approaches to explanation: those emphasizing causation and those emphasizing unification. The causal approach, I shall argue, has a natural account of how and when *ceteris paribus* laws explain. The unification approach fares less well. However, this is not a serious weakness in my argument, for in Section 3.3 I argue that unification is not essential to explanation.

If explanation depends on causation, then *ceteris paribus* laws can explain if they cite causes. Obviously when the *ceteris paribus* clause is satisfied, *ceteris paribus* laws explain because they cite the real causes at work. However, the hard question is how can *ceteris paribus* laws cite operative causes when other things are not equal? They can do so by picking out tendencies.[6] When a *ceteris paribus* law really helps explain, it does so because it identifies one factor in a complex situation. So while other things may not be equal, we can still make sense of the idea that a *ceteris paribus* law is relevant. It covers the phenomena to be explained because it picks out a real aspect or tendency.

Tendencies, in the sense I am using the term, are partial causes, single aspects or factors in a complex causal network. The law of gravity, for example, holds only *ceteris paribus*. Yet it can explain the behavior of subatomic particles even where other factors do interfere. The law of gravity explains in this case because it picks out a component force or cause from the complex causal structure. In my terminology, it picks out a tendency. Defined as "partial causal factors," tendencies are no more mysterious than causation in general. Admittedly, tendencies are frequently associated with much more metaphysical and controversial notions. As I employ them, however, tendencies need not involve those large commitments.

We can clarify tendencies as partial causal factors by looking at what they do *not* require. The law "*A* causes *B, ceteris paribus*" can pick out tendencies without it being the case that *A* is frequently seen to bring about *B*. In other words, a causal law can pick out a tendency – a partial causal factor or influence – even if the tendency itself is never dominant. For example, gravity partly explains the behavior of a body with an electri-

---

[6] This idea has a long pedigree, with John Stuart Mill its most famous early defender. I argued briefly for this idea in Kincaid (1990b). Cartwright (1989) develops and defends the idea in great detail, though her concept of capacities is much more metaphysically robust than my own. Little (1989, p. 201) applies the idea of tendencies to the social sciences.

cal charge. Yet if electrical forces were always greater than gravitational forces, we would never see the body follow the law of gravity. Thus there can be a tendency for *A* to cause *B* without it being the case that *B* happens frequently or at all.

Similarly, tendencies as understood here do not require some notion of approximate truth. The law that *A* causes *B*, *ceteris paribus*, may pick out a real tendency even though it is not even approximately true (whatever that means) that when *A* occurs, *B* does so as well. If other forces always outweigh gravity in some domain, then objects will not even approximately follow the law of gravity. Of course, some *ceteris paribus* laws – sometimes labelled idealizations – are meant to pick out approximate or rough truths. But there is no requirement that they must. Many *ceteris paribus* claims are *abstractions:* they describe particular components from a complex with no claim that the component is dominant.

Tendencies are also not to be equated with counterfactual claims. No doubt *ceteris paribus* laws do entail counterfactual claims – most obviously, that if other things were equal, *A* would result in *B*. But not every counterfactual claim must identify an explanatory tendency.[7] For example, *if* markets were perfect, then economic forces would cause firms to choose the output where marginal cost, marginal revenue and price are equal. However, that counterfactual assertion does not ensure that such a tendency exists in real, imperfect markets. Because other things are not equal, there may be no such tendency at all. In short, true counterfactual claims do not entail true, explanatory tendency claims.

Finally, not every type of *ceteris paribus* law must be analyzed via tendencies. Science produces laws or law-like statements that are apparently not about causes at all. Such laws simply describe behavior. It does not make much sense to think of these as about tendencies. A body moving along a vector may be composed mathematically from two simpler notions. But a behavioral law describing one of those component motions apparently does not describe real tendencies as defended here. If it did, we would be forced to conclude that the planets actually do move in a straight line and fall to the center, not that we can usefully describe their behavior as if they did. Similarly, consider Marx's alleged law that the profit rate tends to fall. If we never observe a decline in profit rates, can we still explain the actual profit trends as resulting from a real tendency of profits to fall? Insofar as Marx's law is a behavioral law rather than a causal one, we cannot. So tendencies may not make much sense for behavioral laws qualified *ceteris paribus*.[8]

---

[7] Here I differ with Gibson (1983).
[8] Cartwright (1989) seems to think the machinery of tendencies and capacities can be extended without problem to behavioral laws.

Is this a problem for taking *ceteris paribus* explanations as being about tendencies? I do not think so. Recall what motivated us to posit tendencies in the first place. It was a desire to show that *ceteris paribus* laws can explain. Yet if we focus on explanation as the citing of causes, then we are not forced to posit tendencies for behavioral laws. Behavioral laws do not cite causes and thus need not explain, whatever their other scientific virtues. Thus tendencies need not lead us to the mysterious conclusion that the planets both move in a straight line and fall to the center.

We can understand how *ceteris paribus* laws explain on the causal approach. It is not so obvious we can do so if we emphasize unification. Laws explain on the unification approach because they take many diverse phenomena and reduce them to an instance of one basic principle. Without an appeal to tendencies, *ceteris paribus* laws are apparently false when other things are not equal. But false statements arguably do not give us real – as opposed to merely possible – explanations. So however much *ceteris paribus* laws unify, they still cannot explain because they are false. *Ceteris paribus* laws remain a mystery on the unification view.

Defenders of the unification approach can either bite the bullet and allow that false statements explain or find their own surrogate for tendencies. The former tack threatens to make explanation a largely psychological, subjective matter. The latter move is not impossible, but proposed notions like "approximate truth" are both hard to clarify and will not suffice for abstractions that are not even close to the truth. Thus I shall not try to defend *ceteris paribus* laws as explaining via unification. Fortunately, we can understand many of them, causally construed.

Given that we can at least make sense of how causal *ceteris paribus* laws explain, we have next to (1) say how we can confirm *ceteris paribus* laws and (2) give criteria for telling when they are irrelevant. The two questions are intimately related. The key issue in both cases is whether data indicates our laws pick out the real causes at work.

Standard scientific methods – especially those designed to separate real causes from spurious ones – will help confirm *ceteris paribus* laws and tell us when those laws explain. Among the testing practices that lend credence to *ceteris paribus* laws are the following:[9]

> (1) We can sometimes show that in some narrow range of cases the *ceteris paribus* conditions are satisfied. Rational economic man is an idealization, but sometimes consumers do act on well-ordered preferences and maximize. Or, we can *make* other things equal

[9] This list expands a set of criteria first proposed by Hausman (1981b, ch. 7).

through controlled experiments.[10] We can then confirm the law directly.

(2) We can sometimes show that although other things are not equal, it makes little difference: the law holds for the large part. Such tests are of course not decisive, since the law might still hold precisely because other things are not equal. However, if we already have independent reason for thinking that interfering factors have a small influence, then showing our law holds approximately is evidence for its truth.

(3) We can look for evidence that the unspecified factors have no systematic influence. Signs of systematic influence raise the prospect that our law holds for spurious reasons; finding only random deviations suggests that we have identified the causes at work. This general strategy is of course one rationale for randomized experimental design and for the requirement in statistical testing that error terms be randomly distributed.

(4) When a *ceteris paribus* law fails to hold in reality, we can nonetheless explain away its failure. Sometimes the counteracting factors can be cited and relevant laws invoked, giving us at least an approximate prediction of their combined effect. Other times the interfering factors may be unique and fall under no known law, yet we can reasonably explain away their influence. Of course, explaining away can become mere ad hoc curve fitting, but it need not; such explanations can and should have independent evidence.

(5) Sometimes we can provide inductive evidence for a *ceteris paribus* law by showing that as conditions approach those required by the *ceteris paribus* clause, the law becomes more predictively accurate – and vice versa. This gives us evidence that our law does not hold accidentally and that the postulated causal process is really producing our data.

(6) We can try to show that our *ceteris paribus* law is what Leamer (1983) calls "sturdy." A *ceteris paribus* law is sturdy if we can add in possible counteracting influences and show that the law still holds. Doing so is an indication that the law describes real factors. In statistical testing, this method involves adding in variables and showing that the alleged relation still holds. Similarly, simulations can give the same result by using background

---

[10] This is Wesley Salmon's (1984, p. 149–150) approach to confirming the counterfactuals implied by causal claims. Most of the methods listed below are devices that try to mimic experimental methods without having literal control of variables.

information to see if possible interfering factors might undercut the law being tested.
(7) We can provide evidence that there exists some mechanism connecting the variables in our purported law. Evidence for a mechanism is evidence that a real causal tendency is at work, even if it is not dominant.
(8) We can sometimes have domain-specific generalizations telling us that specific *ceteris paribus* clauses are no problem. For example, when a biologist uses natural selection to explain traits, he or she makes a *ceteris paribus* assumption that other factors are absent. Over time, biologists have learned when and where that assumption is reasonable. In short, once we use the above methods to confirm *ceteris paribus* claims, we may then infer that other similar *ceteris paribus* clauses are reasonable.
(9) Finally, we can show that a *ceteris paribus* law has other important scientific virtues: for example, that it holds up against diverse kinds of data, that the *ceteris paribus* law, when combined with other well-confirmed generalizations, yields successful predictions, and so on.

When these methods are applied to causal laws, they both help confirm and show that a real tendency is at work. Each increases our confidence that a *ceteris paribus* law fits the facts because it describes a real relation and not because of spurious factors not described.[11]

These procedures are, of course, a more concrete embodiment of the scientific virtues described in Chapter 2. Method (6) – looking for sturdiness – provides both cross tests and fair tests. By seeing if the law holds up when other possible factors are added in, we are testing our law against different background assumptions; if we use the variables of competing theories, then searching for sturdiness is likewise providing a fair test. Looking for more realistic predictions from more realistic qualifications (method [5]) also provides a cross test against different assumptions. Using information from successful tests to infer which qualifications are reliable – method (8) – gives us a cross test of those other results, as does method (9) for assumptions about mechanisms. Methods (3), (4), and (6), which all involve factoring in or out other variables, give independent tests, if those variables are supported by other evidence and not simply motivated by the need to save the generalization at issue.

So the methods listed above for dealing with *ceteris paribus* generaliza-

---

[11] Except method (4), which is obviously compatible with there being no behavioral law at all.

tions flesh out the symptoms of good science described earlier. And, these methods in turn become much more concrete and precise when embodied, for example, in specific statistical techniques. As I argued earlier, broad scientific virtues are abstractions; doing and evaluating science requires successfully realizing these virtues and methods in concrete, domain-specific practices. Of course, there is no fully *mechanical* procedure for applying these methods. Nonetheless, they are powerful tools for establishing causal laws and showing that those laws describe tendencies which explain the phenomena. Exactly when *ceteris paribus* laws are supported and explanatory is a judgment call. But that is the nature of science.

3.2.2 Ceteris paribus *in practice*

So *ceteris paribus* qualifications need not prevent good science. Do they prevent good *social* science? To answer that question, we must ask if the social sciences can and do successfully employ methods like those just sketched. To defend naturalism I obviously must argue that they can. I begin that argument here and extend it in later chapters. This section discusses some good social research in detail, looking at how it deals with the *ceteris paribus* problem. To show that this research not only confirms its *ceteris paribus* generalizations but does so "successfully," I shall compare it to some of the best work in ecology and evolutionary biology. The comparison is favorable – some social research handles the *ceteris paribus* problem as well as some of our best biology.

Jeffrey Paige, in *Agrarian Revolutions* (1975), has produced an exemplary piece of social science research. Building on the Marxist sociological tradition, Paige sets out to determine the primary causes of agrarian political behavior, particularly in developing countries. His work, however, is not doctrinaire Marxism, for he modifies the Marxian view at many places. Not content with anecdotal evidence, Paige produces a sophisticated statistical, cross-national study of agrarian revolutions, revolts and reform movements. Detailed case studies further confirm those results. Paige's results are not beyond criticism. Nonetheless, they are rigorous and careful; they provide compelling evidence that social research can be scientifically respectable. Moreover, Paige explains in terms of large-scale social processes and as a result his work will later be used to defend holism as well. So I want to discuss Paige's results in some detail.

Paige's primary hypothesis is that class structure largely determines agrarian political behavior in developing countries. The social classes in agrarian systems are of two basic kinds, depending on whether they are composed of cultivators or non-cultivators. Cultivators include sharecroppers, resident wage laborers, peasants with small holdings, and usufructuaries; non-cultivators are the landed aristocracy and agricultural corpora-

	Source of Income	
Agrarian class system	Noncultivators	Cultivators
Commercial hacienda	Land	Land (right to cultivate small plots)
Large estate	Land	Wages
Corporation-owned plantation	Capital investment	Wages
Small holding	Capital investment	Land

Figure 1: Paige's four basic agrarian class systems; with the source of income of cultivators and non-cultivators in each system.

tions. These cultivators and non-cultivators in turn fall into different social classes depending on the source of their income, in particular on whether their income comes primarily from land, capital, or wages.

Four basic agrarian class systems are thus possible (see Figure 1). In the first type of system, both cultivators and non-cultivators draw their income primarily from rights to the land rather than from capital or wages. The most common system of this type is the commercial hacienda or manor. It is an individually owned enterprise which does not depend essentially on power-driven processing machinery or other similar capital investments; its workers typically receive compensation by rights to cultivate small plots of land. In the second type of system, non-cultivators draw their income from the land and workers are paid in wages. Usually these systems involve large estates with little or no power-driven machinery and workers who are either sharecroppers or migratory wage laborers. In the third type, non-cultivators draw their income in large part from capital investments and workers are paid in wages. Plantations owned by a commercial corporation typify this sort of system. Crops are processed on site by power-driven machinery and workers are more or less permanent

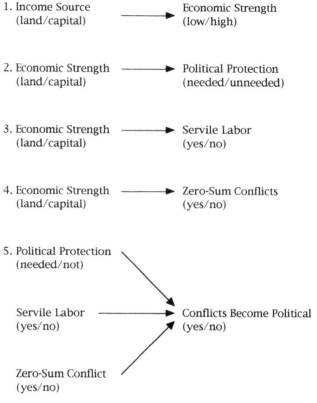

Figure 2: Paige's hypotheses about the influence of income source of non-cultivators on their political behavior.

residents who are paid in money wages. In the fourth type, non-cultivators depend primarily on capital and the cultivators depend primarily on land. Prime examples include small family farms or small-holding peasants producing a cash crop sold to a large agricultural corporation.

Paige predicts that these different economic systems will produce different types of political behavior. To derive specific hypotheses, Paige looks first at how income source affects cultivators and non-cultivators separately. Using those hypotheses he then predicts what happens when those separate behaviors are combined in the four basic class systems. For non-cultivators, Paige proposes the following hypotheses (see Figure 2):

(1) Dependence on land versus capital for income determines the economic strength of non-cultivating classes. Large agricultural

estates relying solely on labor and land are less efficient than other forms of agriculture, especially the small holding. In particular, they are unable to increase productivity by increasing investments in capital. The historical trend shows that the large estate without capital investment loses out in free market competition to other forms.

(2) The economic strength of a non-cultivating class determines whether it will depend on political means to ensure its control of the land. Classes dependent on land are economically pressed; increases in income depend on increases in land holdings. As a result, this agricultural class seeks help from the state to protect its economic position against competition from other forms and encroachments from lower classes wanting land. Classes dependent on capital have much less need for direct political protection since they are economically stronger.

(3) The economic strength of a non-cultivating class also influences the labor it employs. Economically pressed landed estates compete poorly against more efficient forms in free labor markets. This leads the cultivating classes to seek servile labor, and servile labor usually means labor with few political rights as well. Hereditary serfdom and compulsory agricultural labor enforced by colonial regimes are probably the most common forms.

(4) The economic strength of the non-cultivating class also determines whether conflicts with cultivators are zero-sum conflicts. Since classes drawing their income solely from land find productivity increases hard to come by, any dispute with cultivators tends to be a zero-sum game. Upper classes dependent on capital are in a different situation, since they have an increasing pie to divide.

(5) The income source of the upper classes will also determine whether conflicts with other classes tend to be over political authority and property rights or over economic issues and income levels. This hypothesis follows from the earlier ones. The weak economic position of those drawing income from the land makes them dependent on political protection and on servile labor without political rights; any conflict with the cultivating classes is a zero-sum conflict. Thus conflicts over economic issues quickly become conflicts over political control. Those conflicts are (a) not easily resolved and (b) generally outside legal channels, given the servile nature of the cultivators.

Income source also makes an important difference to cultivator behavior. Income from small holdings or from rights to the land lead to quite

74   *Philosophical foundations of the social sciences*

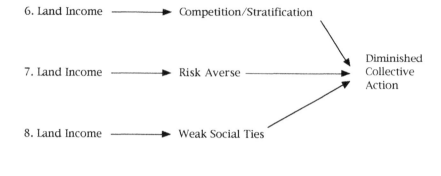

Figure 3: Paige's hypotheses about the influence of income source of cultivators on their political behavior.

distinct incentives and circumstances than does income from wages. Paige (see Figure 3) proposes that:

> (6) Income from rights to land means competitive relations with other cultivators, who are fellow competitors in the market for the crop being produced. Market competition also leads to stratification of cultivators into the relatively rich and poor. As a result of competition and stratification, collective action is diminished.
> (7) Income from land makes cultivators risk averse, especially when it comes to political action. Since the cultivators' land is the only source of income, they have a great deal to lose, and typically begging is the fate of the landless in developing countries. Thus radical political movements will be avoided.
> (8) Income from land means cultivators generally work in isolation and have few inherent ties to other cultivators, leading to weak social ties.
> (9) Income from wages generally indicates cooperative labor by a relatively homogeneous work force who have no property to lose. As a result, collective action, including action over political issues, will be more likely than among those who derive their income from the land.

So, in short, income from land generally undermines collective action, income from wages encourages it.

From these hypotheses Paige is able to predict how different class sys-

Causes, confirmation, and explanation

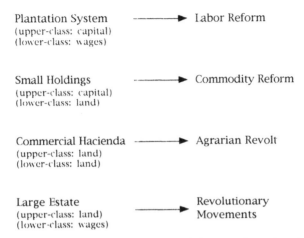

Figure 4: Paige's hypotheses about the influence of agrarian class systems on political behavior.

tems affect agrarian political behavior (see Figure 4). When owners depend upon capital and cultivators on wages (as in large plantations), Paige predicts that cultivators will engage in collective action, but action limited to economic issues such as wages and working conditions and action that ends in compromise settlements. Because of their laboring conditions, cultivators will act collectively over economic issues. Owners, however, are economically strong and generally have an increasing pie to divide; moreover, owners are not seriously dependent on political protection by the state, and their workers can legally act collectively over economic issues. Disputes over economic issues will thus not be naturally transformed into disputes over political power, and compromise will be the name of the game. (Paige thus argues against Marx here.)

The small holding system will likewise produce limited challenges to political authority. Cultivators will draw income from the land and sell products in the market. They will thus be risk averse and divided between rich and poor, reducing the prospects for collective action. If collective action over economic issues occurs, it will not involve challenges to political authority nor be long-lived. The commercial class, which owns the factories processing the products from small holders, does not depend on political protection and is not involved in a zero-sum game. Moreover, the market mediates its relation to small holders, thus minimizing conflict. So small holding systems should result in limited protests over credit, market prices, and the like – what Paige calls a reform commodity movement.

When both cultivators and non-cultivators get their income from the land, we can expect a more severe conflict. Owners are economically weak, dependent upon the state for protection, and unable to compromise by sharing gains in productivity with cultivators. Cultivators are typically without political rights. This combination of factors means economic disputes cannot be easily settled and will naturally spread to issues about land ownership and property redistribution. However, the cultivators are subject to all the factors that undercut collective action. So cultivator movements should not become revolutionary movements, unless other forces like urban political parties intervene to introduce organization from the outside. So, in Paige's terminology, "revolts" may occur, but sustained support for thoroughgoing political revolution should be rare.

Finally, upper-class land income and cultivator income in wages make for the most explosive situation. The upper class is economically pressed, weak, unable to compromise through sharing productivity gains, and dependent on state protection. Cultivators typically are not divided along income lines and have no property to lose in collective action. So economic disputes should be frequent and should quickly become disputes over political authority, since the upper class depends upon political force to maintain itself. Revolutionary movements should thus be most frequent when this class system predominates.

Such are Paige's hypotheses. Are they true? Paige does a meticulous job of arguing that they are. His evidence falls into three basic categories: previous research, mostly case studies, supporting the hypothesized mechanisms linking class structure and political behavior (hypotheses [1]–[9] above); a world study looking primarily for correlations between economic systems and political behavior; and case studies of Peru, Angola and Vietnam. It is the world study which is most original and impressive; since it is also most easily summarized in a short space, I shall concentrate on it here.

Paige divides the developing world into 135 export sectors. He categorizes each sector by the dominant class system and measures agrarian political behavior in each sector. If Paige's hypotheses are right, the data should show that political behavior varies across sectors according to the dominant form of agrarian production. An export sector is a region that is the main producer for a specific export crop within a specific country. Paige picks export sectors as the unit of analysis for several reasons: (a) export sectors, unlike nations, are relatively homogeneous and thus help avoid the problem of spurious correlation facing aggregate data; (b) export sectors tend to dominate the economy of developing nations; (c) it is in export sectors where the landed upper classes are under the most pressure from market forces and thus where Paige's proposed processes should be most visible; and (d) it is mostly in the export sector where several of the

Causes, confirmation, and explanation

class systems – sharecropping and the industrial plantation – are present. Paige does not think that export sectors present the whole story about agrarian political behavior, but he believes focusing on them provides the best test of his hypotheses.

Paige places each of the 135 export sectors into four different categories: (1) the hacienda and (2) the sharecropped or migratory labor estate, both of which involve upper-class income from the land but differ on payment systems for cultivators, and (3) the plantation and (4) the small family holding, both of which involve income from capital. Over the period Paige studied – 1948 to 1970 – nearly all of the 135 sectors fell into one of these categories and nearly all remained in the same category for the entire time data were collected.

The other key component of Paige's data comes from measuring rural political movements in each export sector. Paige does so by counting newspaper reports of agrarian political events – such as land seizures, demonstrations, and so on. For an event to count for the study, it must be collective, outside of established political institutions, involve some sense of shared identity, and be performed by agricultural cultivators. This means panics, mass migrations, and the like were excluded as were events involving primarily students, workers, or urban-based guerrillas. Nonetheless, Paige's data base contained 1,603 events. The coding of these events was checked for reliability by using outside coders; coding was 99 percent reliable.

Paige's theory predicts that events should fall into roughly four different categories. Revolutionary political events were those that involved demands for unconstitutional political change or radical change in rural class structure. Events involving a demand for expropriation and redistribution of land but not unconstitutional political change Paige classifies as instances of agrarian revolt. Labor events are actions around wages, working conditions, and so on but not involving constitutional change or land redistribution. Commodity reform events are protests about the way the market works for agricultural commodities – for example, demands for controls on credit or price supports.

With the data classified into event types and sector types, the test of Paige's hypotheses is obvious: export sectors dominated by commercial hacienda systems should show agrarian revolt; sectors involving sharecropping or migratory labor will show revolutionary activity; in sectors where small holding systems predominate, commodity reform actions will be most common; and where plantations predominate, labor reform events will be found. Using both tests of correlation and path analysis, the data support all of Paige's hypotheses: political behavior is associated with class relations in just the way Paige's theory predicts.

Are Paige's hypotheses qualified *ceteris paribus?* Of course they are. However, he provides relatively good evidence that the *ceteris paribus* qualification is no fatal flaw. For starters, we know that Paige's results do hold for the most part regardless of disturbing influences, because he finds a strong positive correlation between the proposed event types and sector types. Paige also shows that his results can be refined; adding in complicating factors does not undermine his results and in fact generally strengthens them. For example, Paige considers the following complicating variables:

(1) *contagion effects:* The probability of an event occurring at one period may be considerably increased simply by the fact that such an event occurred in the previous period. We can imagine numerous mechanisms that might make contagion effects likely – success at one time makes recruitment to a movement easier at subsequent times, publicity from one event increases support for similar events thereafter, and so on. To be sure he is measuring the influence of class structure rather than contagion effects, Paige uses an exponential function to describe the number of events. Since contagion effects are likely to mimic an exponential growth function, using the latter is a way of controlling for contagion effects.

(2) *centralized versus decentralized production:* We can break sharecropping systems into two kinds, depending on how directly the landowner controls work. Centralized systems, which typically are cotton estates, actually have much in common with the commercial hacienda or manor. Plots in centralized sharecropping are often inherited; workers are frequently deeply in debt to a company store owned by the landlord; and long-term tenure is common. These characteristics make sharecroppers in the centralized system dependent on the landlord and risk averse, much like cultivators in the commercial hacienda. Paige controls for this complication by subdividing the sharecropping sectors into decentralized and centralized production. He finds that doing so increases the strength of the predicted relationship.

(3) *the presence of urban parties:* Paige predicted that peasants working on commercial haciendas would be involved in agrarian revolts over land, but that those revolts were likely to be short-lived and unsuccessful because the peasants' economic circumstances undermined collective action. However, if organization is introduced from the outside in the form of urban political parties, then we would expect the odds of agrarian revolts to increase. So Paige adds in a variable for the presence of urban revolutionary parties

Causes, confirmation, and explanation

to his tests and finds that (a) the predicted correlation increases in strength and (b) the correlation between agrarian revolt and the commercial hacienda system is not due simply to the presence of urban parties.

(4) *extent of the market:* Small family holdings differ considerably. Some are only minimally involved in producing for a market and involve primarily subsistence farming. Thus Paige's prediction that the small family holding system will lead to commodity reform movements should get more accurate if we factor in the degree of market participation. Paige uses a measure of market participation for each sector to test this refined hypothesis; he finds that the original correlation increases in strength as predicted.

So Paige identifies numerous factors falling under the *ceteris paribus* clause and shows that his causal claims hold up and are strengthened when these factors are included.

Paige also explains away the most obvious exceptions to his hypothesis. For example, Malayan agricultural wage earners were involved in revolutionary movements, yet they should not have been on Paige's theory. Paige explains away this apparent exception without invoking ad hoc devices. The Malayan cultivators in question worked on rubber estates. Rubber production involves conditions very different from the typical industrial plantation. It does not require large quantities of capital; economies of scale are minimal compared to other crops. As a result, owners are in an economic situation much closer to that of the typical landed estate, where income is derived from land, not capital. Improvements in productivity are limited. Thus the Malayan rubber plantations are similar to the landed estates which do tend to produce revolutionary movements. So the Malayan case, once the details are filled in, fits the predictions of the theory rather than contradicting them.

Finally, we noted in the last section that identifying reasonable mechanisms can help support *ceteris paribus* claims. Paige has also done that. Hypotheses (1)–(9) are the mechanisms tying class relations to political behavior. Those mechanisms had much prior plausibility from previous studies, and Paige himself also provides some new evidence for them as well. In particular, he provides indirect evidence that commercial haciendas are inefficient, a key postulated mechanism.

So Paige's theory of agrarian movements certainly seems to have the traits of good science. It exhibits the *evidential* virtues summarized by independent, fair, and cross tests. Though its laws are qualified *ceteris paribus,* it confirms those claims by applying testing methods common in the natural sciences. Paige's theory also seems to have explanatory virtues: it explains by providing relevant *causal generalizations*. Those causal generalizations, of course, describe tendencies or partial causal factors. Yet Paige's testing procedures provide good evidence that those tendencies are actually operative. Paige's work thus gives us a prima facie case that good science in the social sciences is not only possible but sometimes actual.

Skeptics may grant my claims so far but remain unconvinced. Even if Paige's work does embody basic scientific virtues, it may not do so *well enough*. To assuage such doubts, I want to argue next that Paige's work compares favorably with much good work in the natural sciences. I do so by looking at evolutionary theory and ecology and arguing that they face roughly the same problems with the same success as does Paige. The end result will be a yet stronger case for naturalism.

I start with ecology. Modern ecology roughly divides into two domains: community ecology and population ecology. Community ecology describes the overarching structure and function of concrete ecosystems; population ecology searches for the general principles that determine the rise and fall of individual populations. Obviously the two domains are interrelated.

Ecologists have made continual, albeit slow, progress in identifying the main factors determining population dynamics.[12] In the broadest terms, population dynamics depend upon the relative rates of emigration and immigration and birth and death. Behind these factors lie numerous complex determinants. Each organism has reproductive potential. Each environment has carrying capacity. Environment and reproductive potential interact to determine population size. These two factors can again be broken down. Understanding how a specific population develops requires understanding exactly what in the environment causes organisms to die. It also requires knowing at what part of the life cycle those factors operate and how they depend upon the traits of individual organisms. Population ecology is thus naturally led to consider species interaction – for example, predator–prey interactions. These are in turn explained in part by the search strategies of predators and avoidance strategies of prey. Obviously things get complex very fast. But population ecologists have nonetheless developed general models that describe how these factors interact and that allow them to identify, at least for some specific populations, the factors determining population dynamics.

Despite these successes, population ecology has much to learn. While they can identify major causes or limiting factors, ecologists can neither cite all the variables nor determine with precision how known variables

---

[12] Most of what follows can be found in any text on population ecology. See, for example, Berryman (1981).

interact. The influence of ecological variables is often context sensitive: effects differ as population density differs and as time lags change. Understanding how predation and disease affect population in one concrete ecosystem at one population density does not guarantee we understand how those factors work elsewhere. Density dependence and time lags are not of course the only complexities. Variable environments and patchy habitats may make important differences as well.

Thus population ecology really looks much like the best work in the social sciences. We can cite with some certainty major mechanisms and dynamics. Some phenomena are understood in detail. But most studies make simplifying assumptions and leave variables uncontrolled. Even when we know the variables, we often have no general, automatic way to combine them or describe their interaction: putting causal factors together is often a piecemeal and ad hoc process. We can make generalizations, but they are bound to have exceptions and to assume that other variables, known and unknown, do not interfere. Also, most generalizations hold only for specific kinds of populations.

Community ecology is in much the same boat. It describes, for example, laws of succession. Such laws are known to have exceptions; among the counteracting factors is radical environmental change, though other unknown factors may be involved. Similarly, another prime law from community ecology – namely, that organisms with the same ecologies cannot coexist in the same environment – may fail to hold when the environment is sufficiently variable, when population sizes at equilibrium are equal, and when selective forces are weak. The complexities of real competitive situations make it unlikely that these are the only possible counteracting factors.

Ecology is no exception. Evolutionary theory, which is the heart of biology, the theory that ties all of biology together, looks much the same. The basic components of evolutionary theory are of course well developed. The Hardy-Weinberg law describes gene frequencies in a population when no forces are present. Natural selection acting on variation causes changes from equilibrium. Theoretically, selective forces can be quantified and resulting trajectories of populations determined for particular situations. For simple selective forces acting on individuals of known fitness, the results are known with mathematical precision.

However, selection is not the only force acting on real populations. Under certain conditions – for example, small population size or weak selection – gene frequencies may change due to random factors. The result is called genetic drift, and it too can be factored in by population genetics. Selection also involves more than the mortalities and fecundity of individuals. Individuals low on both measures may nonetheless be well represented in the population, due to what is generally called kin selection. An organism may spread its genes by ways other than ensuring its own survival or that of its own offspring. Siblings and others may also carry an organism's genes. Helping them survive is thus an indirect contribution to fitness. So effects on kin also have to be incorporated into the basic theory. Finally, organisms may survive long and have great reproductive potential – and yet be relatively unfit. Mates must be attracted. So some traits may be selected for sexual reasons, not because they contribute to adaptedness in the normal sense. Thus "sexual selection" must also be added to any complete model.

These additional factors suggest that evolutionary explanations will be qualified *ceteris paribus* and that adding in interfering factors will be a complex process much like that in ecology. However, drift, sexual selection, and kin selection are just the beginning. Numerous other factors influence real populations. Selection is standardly taken to operate on individuals. Yet selection can operate at both higher and lower levels – on genes and chromosomes and on trait groups, populations and perhaps even species or species groups. Thus selection at multiple levels may be a complicating factor. Models exist to incorporate some multi-level forces, but even on the theoretical level biologists are not sure how exactly to combine them with more ordinary determinants of evolution.

Similar complexity arises from other known influences on evolution. Basic population genetic models treat genes like beads on a string – they are independent and separable. The real world is far more complex. Genes interact to produce a given trait or produce multiple traits. Such "gene linkage" throws off predictions made from simpler models. Furthermore, basic evolutionary theory often assumes that natural selection will produce those traits best promoting fitness. Again, the real world is more complex. Developmental constraints may influence evolution in important ways. We have at best only a rough idea how developmental constraints are to be factored in; usually, we are limited to citing their qualitative effects in particular cases.[13]

The complexities do not stop here. Talk of a selective "force" is misleading and metaphorical. In reality, organisms face a complex environment influencing their survival in numerous complex ways. Environments are often "patchy," and there is no automatic, theoretical way to factor in such variability. Furthermore, any real account of selective forces inherits ecology's complexity as well, for it is ecological factors that make up the environment. Finally, not all selection has to be genetically based. Organisms do learn. Those that learn well and pass that learning on will survive

---

[13] For a discussion of developmental constraints, see Maynard Smith et al. (1985).

more frequently. Explaining behavior thus may require including the effects of cultural selection.

Evolutionary theory thus paints a familiar picture: We know the major factors and how they interact. We can control for some interfering factors, but we must ignore or abstract from others. Some factors we know how to incorporate in a theoretical way, others we do not. Causal generalizations are bound to be exception prone and to hold only for restricted domains.

These complications confront even the most careful evolutionary research, as we can see by considering a paradigmatic piece of good evolutionary research. Peter Grant's *Ecology and Evolution of Darwin's Finches* (1986) is the definitive research to date on finch evolution on the Galápagos Islands. It is widely recognized as a paradigm of evolutionary research. Nonetheless, Grant's work runs into all the complications discussed above. I want to discuss Grant's results in some detail, for they nicely illustrate just how similar Paige's work is to good work in the natural sciences.

Grant's task is to describe the evolutionary forces that produced the remarkable diversity of finches in the Galápagos. From more than ten years of study in the islands plus a mass of previous work, Grant develops a detailed evolutionary picture. Finch beaks are adapted to different seed sizes. Seed availability is a strong constraint on population size. Selection is currently occurring among some populations, favoring large body and beak size. Current species apparently originated under gradual directional selection. Speciation was driven by both allopatry (separation with subsequent adaptation to different food sources) and competitive exclusion.

Grant recognizes the many possible complicating factors and attempts to rule them out. Drift seems not to be important, since finches flock and populations are usually relatively large. Internal constraints appear minimal, since beak and body size involve small, accumulative genetic changes. Nonetheless, Grant's work faces numerous complications. His predictions are qualitative – he can argue for the direction of selection, but not its strength. Mutation rates and gene flow are complicating factors, but their relative importance is unknown. Sexual selection has played some role, but it is not clear how much nor how exactly (1986, p. 103). Rain, large seed production, timing of seed production, and other food sources may influence to what extent seed actually serves as a population limit and selective force; at best, Grant gives reasons to think these factors are less important. Predation and disease are additional selective forces that are not included in his model; again, field experience indicates they are seldom important. Competition and habitat diversity play an unclear role in speciation. Hybridization does occur, but the simplest explanation is one that gives it small place. While selection clearly picks out beak size, it is not clear *how* it does so – the precise mechanism is uncertain. Finally, Grant's major findings have exceptions in the Galápagos themselves, exceptions that must be explained away by appeal to unusual circumstances.

My point is not to challenge the quality of Grant's work. Far from it. The correct inference is rather that good work in the social sciences faces problems very much like those confronting the best work in biology and handles them with the same approximate success. *Ceteris paribus* generalizations and the ensuing complexities do not separate the social sciences from the natural sciences.

3.3 Inferring causes from non-experimental data

The previous section made a prima facie case that the social sciences can meet standards of scientific adequacy. Some critics may still think that my case is weak at best. After all, Paige's work differs from good natural science in at least three ways: (1) Paige's evidence is entirely non-experimental, (2) it relies on establishing causation simply from correlation, and (3) it uses narrative case studies as evidence. Yet good science seems to directly establish causation by manipulating the world; it surely does not rely on impressionistic discussions of particular events. Some of these worries are more serious than others. All can be heard time and again from critics of the social sciences. This section addresses these more general doubts about confirmation in the social sciences.

Controlled experiments are not unknown in the social sciences. Some occur in economics (see Loomes 1989). Psychology of course also employs them. Yet, experimental evidence occurs only in a small fraction of social science research. Barring the complete elimination of the social sciences for experimental psychology or neurobiology, social research must be partly non-experimental. So challenges to non-experimental evidence are serious problems.

A first reply to this worry is that non-experimental evidence is also rife in the *natural* sciences. Darwin's support for the theory of natural selection was almost entirely non-experimental; as the work of Grant shows, current evidence in evolutionary biology is largely non-experimental. In ecology, experiments do occur, but they pale in importance compared to purely observational, descriptive evidence. Moreover, non-experimental evidence is central to physics. General relativity rested on non-experimental evidence for most of this century; much of contemporary astronomy does so as well. So if only experiments confirm, some of our best natural science is inadequate as well.

Aside from this "physics does it too" argument, we have good philosophical grounds for thinking experimental evidence is not a sine qua non for good science. Chapter 2 sketched Quine's doctrine that testing is holistic. Every experimental test presupposes numerous auxiliary assumptions. Those assumptions ensure that we are holding everything else constant when we wiggle the variable of interest. As a result, experimental evidence does not guarantee certainty – our outcome is only as good as the assumptions we presuppose. Many apparently well-designed experiments fail to test their hypotheses, because relevant variables are ignored, because the experimental setup does not measure what the experimenter thought, and so on. So experimental data is fallible. It confirms only when we have made the right background assumptions. However, "natural experiments" – evidence from non-experimental observations – are in the same boat: if the right background assumptions hold, they can provide compelling evidence.[14] Thus experimental and non-experimental tests have the same basic structure: using what we already know along with what we observe to draw conclusions.

If there is no difference in principle, there is of course one in practice. Experimental data make it generally easier to eliminate spurious causes and test hypotheses. Experiments allow us literally to control interfering factors. Natural experiments have no such luxury. This practical obstacle shows up forcefully in Paige's data. That work, like much of the best social science, tried to infer causation by applying statistical inference to non-experimental data. Yet, inferring causation from facts about association or correlation is risky business. The problems are many, but let me mention two. At the most fundamental level, causation apparently cannot be fully analyzed in terms of regularities and probabilities. Though some philosophers would disagree, compelling arguments show that causal notions cannot simply be replaced by probabilistic ones (Cartwright 1989). This suggests that information solely about probabilities or correlations will never suffice to establish causes. So statistical inference alone will not guarantee causes from non-experimental data.

More concrete problems come from the actual practice of statistical inference. Paige provided a causal model of agrarian behavior and tried to infer causes from statistical correlations. As Glymour et al. (1987) have pointed out, causal relations between even a few variables can potentially be modeled in many different ways. For example, for four causal factors where mutual causation is a possibility, the number of potential models is 4^6, or 256. As a result, finding that one particular model is consistent with the statistical correlations may be inconclusive proof. Good science does fair tests, and fair tests do not rule out competing explanations by fiat. However, until we show that our favored causal story is better than the many possible alternatives, we are doing precisely that.

---

[14] See Cartwright (1989) for a detailed defense of this claim.

These complaints are legitimate. However, at most they demonstrate that good social research is hard – and harder than *some* research in the natural sciences. Why "some"? Causal inference from non-experimental correlations is not unique to the social sciences. Evolutionary biology and ecology rely extensively on statistical inferences to causes from nonexperimental data. So we should immediately suspect that these difficulties are also not principled obstacles to naturalism. Good science finds ways to infer causes from statistics despite the difficulties.

Inferring causes from correlations is hard – or too easy – because correlations are thin evidence. To tie down actual causes from observed correlations we need further constraints. Those constraints can be either *logical* or *causal*. Logical constraints involve other, more complex statistical relations implied by the model at issue. Simon's (1971) earlier work on causal ordering and the current work of Glymour et al. (1987) illustrate such an approach. Causal constraints, on the other hand, allow us to use background causal information to go from correlations to causes. Cartwright (1989) argues that logical constraints by themselves do not suffice; she also argues that *proving* causality from correlations requires quite stringent assumptions.

For my purposes here, I can remove these worries by arguing two points. First, while inferring causes from non-experimental correlations is hard, so is inferring causes from experimental data. To *prove* causation experimentally we must know all possible relevant background causes and control them all. Few, if any, real experiments meet this requirement. Scientists sometimes try to avoid this problem by using randomized experimental design. Yet randomization is no panacea.[15] Randomized experiments can give the wrong answer in complex situations. If a particular treatment has dual capacities – say causing X in some individuals and preventing X in others – then a randomized design may mistakenly find no causal influence. Such problems are avoided by breaking up the treatment group into subgroups composed of individuals where X is produced and those where it is prevented. Doing that, however, presupposes a good deal of background information about how the cause works and so on. So randomized experimental design by itself is no guarantee of believable results. The moral is that inference to causes is always hard, inside natural science and out.

My second response to worries about causal inference in the social sciences is simply that some social research makes such inferences reasonably well. Paige again provides a good example. One of Paige's causal models

---

[15] For doubts that randomization has any special epistemic rationale, see Urbach (1985).

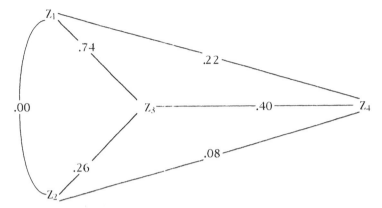

Figure 5: Diagram of the correlations found by Paige between commercial hacienda (Z_1), reform parties (Z_2), their interaction (Z_3), and agrarian events (Z_4).

concerns the effect of the commercial hacienda system on agrarian behavior. Paige claims that the commercial hacienda system promotes agrarian rebellion over distributional issues, but not revolutionary demands. Since both owners and cultivators draw their income from land, not capital, conflict is inevitable; since peasants are isolated and risk averse and have competing interests, they are unlikely to engage in persistent collective action for radical demands. As we saw earlier, this generalization holds only *ceteris paribus*. Among the interfering factors are urban political parties. Urban parties may introduce organization from the outside and turn isolated and sporadic protest into a movement for agrarian reform.

Paige thus proposes the causal model depicted in Figure 5, which lists correlations produced by statistical path analysis. This is only one of many possible *a priori* models. However, Paige has a good case that this is the best model for the data. The commercial hacienda was a constant in the period Paige studied, making it unlikely that this basic class structure came to exist because of agrarian movements. Thus a causal connection from Z_4 to Z_3 is ruled out. The correlation between agrarian movements and socialist parties (the Z_2-to-Z_4 path) is very weak. We also know that socialist parties have their roots in the urban population. Hence any real causal path from Z_4 to Z_2 is unlikely. Furthermore, the commercial hacienda system and regions with socialist parties are not correlated, so Z_1 does not cause Z_2 or vice versa. Since two effects of a common cause would show a correlation, Z_1 and Z_2 cannot be produced by their joint interaction effect (if that even makes sense). Finally, as Paige claims, the interaction effect accounts for most of the small direct correlation between commercial haciendas and agrarian movements. Thus, between background causal information and the facts about the correlations, we can seemingly rule out most alternative causal models.

No doubt Paige's work is far better than much social research. In particular, Paige's *ceteris paribus* clauses are refined and his initial variables proposed on theoretical grounds. Unlike much social research, Paige's generalizations are not the result of piecemeal curve fitting. Still, however atypical Paige's work may be, it gives us good evidence that the problems of non-experimental evidence need not prevent science in the social sciences. After all, Paige's research is not on some isolated and unusually simple social phenomena – it is about something complicated and fundamental, namely, the connection between social class and political behavior. What Paige achieves can be accomplished elsewhere in the social sciences.

Another worry about causal inference in social research concerns the holistic or macrosociological nature of its evidence. Paige's claims are about large-scale entities; his regressions relate aggregate variables. Some may doubt that aggregates can be real causal factors or that aggregate or average data suffice for causal inference. While these worries are legitimate, I mention them only to postpone their discussion. These doubts are about the holistic nature of social theory. We take up that second big issue in Chapter 5 and in the process we shall consider (and reject) claims that macrolevel causation and data are inadequate.

So far I have addressed two of the three basic worries mentioned at the beginning of this section. The third complaint concerned case study evidence. Paige used case studies; detailed examination of individual cases is standard fare in the social sciences, especially anthropology. This fact raises doubts about naturalism on two grounds. First, it appears to make social science evidence very different from evidence in the natural sciences. If the replicable experiment exemplifies good science, case studies seem just the opposite: they describe unique circumstances and unrepeatable evidence. Since case studies select from some subset of the possible facts, they likewise seem inherently biased. Furthermore, if science is about producing laws, then case studies seem precisely the wrong sort of evidence. Case studies are about unique circumstances; laws are about what is universal. The old debate about nomothetic versus ideographic pictures of the social sciences turned on precisely such concerns.

These worries point to real difficulties for good social scientific practice. Yet, as with other objections considered in this chapter, these difficulties are not insurmountable obstacles. Each can be handled, in principle and in practice.

As every experimental scientist knows, "replication" comes in degrees. Most experiments involve such extensive background knowledge and skill that exact replication is impossible. Rather, good science hopes to replicate previous work "well enough" and in the relevant respects for evaluating the hypothesis at issue. Furthermore, social science case studies *can* be replicated in some ways. Paige describes the events and facts surrounding peasant movements in Vietnam; many of these facts are quantitative. His claims are consequently subject to scrutiny by standard methods. And, even a participant observer case study in a remote tribe can be double-checked after a fashion. Margaret Mead's work on Samoa, for example, was re-examined and largely rejected for failing to stand up to the evidence (Freeman 1983).

Case study evidence surely does present only a slice of the possible data. Nonetheless, this problem is endemic to science, not unique to case study evidence. No experiment reports every possible observation. The crucial question rather is whether the data cited represent a fair test. A fair test makes no background assumptions that bias the case against one competing explanation over another. It is no doubt much easier in a case study to hide potentially troublesome facts and report only those that confirm the hypothesis advanced. Yet, such bias is not inevitable.

It is also true that case studies may not represent the larger population. However, for some purposes that may be irrelevant, since we may not be interested in the larger case. The facts about evolution on Earth may be entirely unrepresentative of change in other life forms on other planets. But for those with less cosmic concerns, that hardly matters. Furthermore, representativeness is a contextual virtue, valuable in some situations, less so in others. When we want to make precise inferences about large populations from small samples, representativeness is paramount. However, we are not always in such a situation. Sometimes lower standards will do. In epidemiology, for example, researchers frequently study risk factors in a major urban area. They make no pretense that the city studied is a perfectly representative sample of the nation at large. Usually, however, we have no good reason to think that case is highly atypical. In short, we use our background knowledge to judge that the case being investigated is typical enough.

Paige, for example, uses three case studies to help support his theory. He believes that these cases are typical of agrarian movements in general. Where special features intervene, he tries to account for their influence and factor in the difference they make. Paige may be wrong. But the key point is that we can use cross checks to see if his case studies are representative.

Case studies can also tell us something beyond the particular. While case studies are of single events, those events inevitably have multiple aspects. So even within a case study we may find multiple instances to test generalizations (see Campbell 1975). Paige's study of Vietnam, for example, provides multiple tests for his hypotheses about peasant behavior – tests about differences in behavior in different regions of the country, about change in behavior as the economy and political parties changed, and so on. Case studies can also help rule out competing hypotheses and thus help confirm causal explanations in the processes. There may be no "crucial" tests in science, if a crucial test is one that cannot be explained away by changing background assumptions. However, single tests can be decisive evidence when they are fair tests – when the test relies on background assumptions shared by the main competing theories.

So where do case studies stand? As with the other objections in this section, I have shown that the problem is not a principled one, that the problem shows up in the natural sciences and that social scientists can sometimes handle the problem in practice. Of course, if the social sciences only had case study evidence, there would be serious doubts about naturalism (though arguably Darwin had little else either). But that is not true.

Non-experimental data, causal inference from correlations, and case study evidence do not make the social sciences inevitably pseudoscientific. A more compelling case for naturalism would look at research based on these kinds of data in more detail than we have here. Still, the arguments of this section and the previous one strongly suggest that good evidence in the social sciences is possible in principle and sometimes in practice.

3.4 Lawless explanations

Critics may feel that the previous sections, even if successful, have not shown anything terribly significant. At most, the complaint might run, I have shown that some social research can produce rough low-level generalizations. However, our best science produces much more: laws that are universal and embedded in systematic theories. Only then do we have the unifying explanations that characterize good science. Piecemeal, exception-ridden generalizations are not enough for good science. In this last section I answer these worries and in the process discuss the nature of explanation in the social sciences.

Generalizations with restricted scope and known and unknown exceptions are rife in our best biology, as we saw in earlier sections. By itself this fact casts serious doubt upon the assertion that well-confirmed causal generalizations are not enough for good science. It also raises questions about what justifies requirements for laws. Philosophers have typically looked for formal criteria – criteria that do not rest on substantive empirical detail – for lawfulness. As our Quinean naturalism would expect, that search has not succeeded. Syntactic universality, for example, does not do the job. A sentence that contains no names or definite descriptions may nonetheless implicitly refer to particulars, depending on how its other terms are interpreted; sentences referring to particular entities can be reformulated without them (laws describing our solar system can be treated as laws about any Keplerian system). Sometimes it is said that laws, unlike generalizations, warrant counterfactuals – statements about what would happen under a wide variety of possible circumstances. Yet knowing what circumstances are possible requires that we already know the laws of nature (see Salmon 1992); moreover, low-level generalizations can entail counterfactuals in some circumstances. Another intuitive criterion for lawfulness – holding in all space-time regions – does not suffice either. It entails that most apparent laws in natural sciences outside physics are not laws at all, since they are not space-time invariant. Furthermore, we cannot apply this criterion without prior knowledge of the laws that determine what space-times are possible. Once again, purely formal criteria seem not to suffice.[16]

The moral is that lawfulness is a substantive issue, one that philosophers are not likely to decide by analyzing necessary and sufficient conditions. Nelson Goodman (1965) describes the substantive issues in terms of "projectability" and "entrenchment"; Brian Skyrms (1980), in terms of "resilience." These notions roughly concern how well a given generalization fits with what else we know – how central it is to other claims we make, how probable it is relative to our other beliefs, and so on. There are two important things to note about these accounts. First, these criteria are much the same as those we proposed for evaluating *ceteris paribus* claims. So if the social sciences can produce relatively well-confirmed *ceteris paribus* generalizations, those generalizations will be relatively law-like. Of course I have argued at length that the social sciences can do precisely that.

Second, accounts like those of Goodman and Skyrms leave lawfulness, or at least estimates of it, a matter of degree. If, however, there is no sharp dividing line between laws and well-entrenched generalizations, then the law–generalization distinction cannot carry much weight. Confirming causes, causal generalizations, and laws of universal scope all result from a common process, namely, fair, independent, and cross tests. They also largely aim, I would suggest, at telling us the same thing, namely, the causal structure of the world. So the law–generalization distinction seems a misguided ground for separating the natural and social sciences.

---

[16] See Earman (1978) for a helpful discussion of this last criterion. Earman defends "true in all space-time regions," but simply swallows the conclusion that most laws in the special sciences are not laws at all.

Even if generalizations explain only if they are universal, I think we can make a good case that Paige's work passes the test. As we saw above, universality is a matter of degree and thus critics would have to show that Paige's causal generalizations are not *sufficiently* universal or unifying. Since we have no detailed account of universality, it is hard to make such an argument. However, it seems to me that on intuitive grounds Paige's generalizations are relatively universal in that they:

(1) refer to apparent *kinds:* for example, commercial hacienda, migrant workers, landed aristocracy, and so on.
(2) hold for an unrestricted population: laws are insufficiently universal when their domain of application is arbitrarily restricted. However, Paige's theory has as its domain the entire world population of agrarian societies.
(3) are resilient: as we saw above, Paige's claims persist when interfering elements are factored in.
(4) handle complicating factors in a theoretical way. Many of the *ceteris paribus* qualifications are in fact predicted by Paige's basic theory.
(5) are grounded on reasonable mechanisms.
(6) bring under one theory a set of phenomena previously given divergent explanations: earlier accounts of agrarian behavior had invoked multiple causal processes where Paige identifies one.

Paige's generalizations do look explanatory even if we assume that explanation requires universality and unification. However, much social research does provide piecemeal or singular causal accounts without any very general overarching theory. Thus critics may continue to think that Paige's work is an exception. I reply to this worry in several ways. First, the next chapter on functionalism describes further social research that employs a relatively general theory; Chapter 7 argues that economics likewise provides some well-confirmed models that unify a diversity of phenomena. So theory-driven, unifying accounts occur elsewhere in the social sciences. Second, unifying and unified theories are not essential to good science; it is the citing of causes that counts. I turn now to defend this second claim and to sketch in the process what seems to me a better picture of explanation in the social sciences.

A first problem is that "unification" is an extremely fuzzy notion. Philip Kitcher (1989) has made admirable attempts to expand the intuitive idea. According to him, a theory unifies if it uses a single argument pattern to provide the best systemization of our beliefs. Kitcher expands this idea with some rigor. However, it is not clear we still have any very strong grip on unification. The problems are multiple. For starters, we would like to know when two patterns are the same and when they differ. Kitcher, for example, distinguishes multiple argument patterns in evolutionary biology: simple selection, directional selection, and so on. Why are these many patterns rather than one, more complex pattern? When is one pattern simpler than another? These are not trivial questions, for judgments about unification depend on how we answer it. Relatedly, we need some account of how the various factors in Kitcher's account are traded off. How do we weigh the simplicity of a pattern against the scope of conclusions it produces? We must do so to understand what the "best systemization of our beliefs" involves. I do not claim that these problems are insurmountable. Yet they illustrate how hard it is to give any precise account of unification.

A second worry is that unification is too *subjective*. Explanation is frequently tied to understanding – a good explanation shows that a phenomenon was expected or makes sense. Hempel sometimes slipped into this way of talking. Kitcher's argument patterns are one attempt to cash out this notion; we better understand when we see how phenomena fit into a pattern. But what we "expect" or "make sense of" threatens to become a mere psychological fact about us, not a fact about the world. Doubts raised above about specifying the same pattern, the simplest pattern, or the best systemization illustrate this danger. If unification can be done in multiple ways, then it looks like what explains is not a factual matter about the world but a subjective one about us. Science as an objective enterprise threatens to go out the window.

Unification bears a strong resemblance to another controversial scientific virtue, simplicity. Simplicity has always seemed an odd reason to think one theory more probable than another, because why assume the world is always simple? The same question also confronts unification: Why assume the world is unified? This is no idle question. In fact, I shall argue below and in later chapters that diversity of processes is the rule, not the exception, outside physics. The social and biological sciences deal with processes, mechanisms, and causes that are often highly specific and even unique. Underlying mechanisms vary greatly. So a unified story is sometimes likely to be a false story.[17]

Kitcher's solution to this problem is to argue that when all the evidence

---

[17] Defenders of unification can of course deny that explanatory virtues do give us any particular reason to think theories are *true*. I agree that inference to the best explanation has often been abused (see Day and Kincaid 1994), but the prospect that explanatory power is good evidence that a theory is false seems to go too far in the other direction.

is in at the end of inquiry, every confirmed causal claim will be one that contributes to unification. Causal claims are thus ultimately derivative of unification. This defense, reminiscent of Kant and Peirce, is of course only a promissory note. Appeals to "the end of inquiry" certainly put the burden of proof on Kitcher. Insofar as causation at this point of inquiry seems not essentially tied to unification, that burden is a large one to bear.

Where unification does seem important, we can account for its importance in other ways. More specifically, unity frequently gains its warrant from its role in either confirmation, pragmatics, or causal explanation. Unified accounts frequently promote *evidential* virtues rather than explanatory ones. We may know from background information that probably some single mechanism is at work in a given domain. Then theories that propose one common process are better explanations *in that they are better confirmed*. Similarly, a unified theory probably has survived more and diverse cross tests. Again, unification here is an evidential virtue, for it gives us more reason to think our account *true*. But the truth of a theory and whether it explains well are two different virtues. Thus unity can play an important role in science without being a necessary condition for explanation. Defenders of the unification approach have generally ignored this role for unification – thus making it falsely seem that other approaches must throw out unity altogether (for example, Kitcher 1989, p. 497; Woodward 1989, p. 367).

We can give unity a non-explanatory role in other ways as well. Kitcher's argument patterns have obvious heuristic functions. For a defender of the causal view, his argument patterns look like useful simplifications or guidelines for finding the most likely *causes*. Thus Darwinian biology looks for selective explanations first. This is a strategy for pursuit – for investigating the causes of biological diversity. Furthermore, unity can be the by-product of causal explanation. What made Priestley's and Dalton's law of fixed proportions explanatory? Initially, we might argue that it did not explain, only described. That account became explanatory when it was conjoined with a common underlying causal mechanism, namely, atomic structure. Again, we need not equate explanation with unification to give the latter a place.

Finally, the unification approach simply does not fit much good science. I can make this point with theories from biology we have already discussed. Kitcher claims that evolutionary biology supports the unification account, but our earlier discussion makes that unlikely. The basic generalizations of evolutionary biology are strongly qualified *ceteris paribus*. As we saw, actual evolutionary explanation works by adding in numerous outside factors, often on a piecemeal and ad hoc basis. For Grant to explain

Darwin's finches, he must incorporate factors other than simple selection. He must do so because he is trying to explain the *causes* of current traits. The final explanation is a complex and unique list of relevant causal factors, some selective and some not.

Kitcher implicitly recognizes this fact, for he identifies three different patterns corresponding to the different kinds of selective processes. However, we saw in the last section that there are many more such causal factors potentially at work. Do they too have their own argument patterns? Maybe so, but then unification seems to be replaced by a diversity of complex causal explanations. Those explanations can sometimes be summarized by simpler abstractions, but those abstractions are literally false. The full explanation comes when we identify the long list of relevant causal factors.

We can make a similar argument for another and "harder" biological science: cell or molecular biology. Cell biology is the epitome of a science where causes are paramount, unification scarce. While we can generalize about cellular processes, those generalizations are full of exceptions. There are exceptions because natural selection cares about consequences, not means. So long as fitness is enhanced, the lower-level mechanism causing fitness is irrelevant. That means general cellular processes are frequently brought about in diverse ways. To describe general processes we have to abstract from that diversity of detail. Those abstractions may be useful, but they are derivative of the specific causal processes they subsume. Sometimes sufficient underlying mechanism is conserved – as happens roughly with DNA – to ensure that a tight generalization is possible. But just as often diversity rules. In either case, explanation comes from identifying the specific operative processes.

Doesn't cell biology unify by simply applying biochemistry? Biochemistry has an important role to play, but not the one we would expect from the unification approach. For one, biochemistry is only a part of explanations in cell biology. As I have argued elsewhere (Kincaid 1990d), biological explanations have an apparently ineliminable role, largely because there is such diversity of underlying mechanisms. When biochemistry does come into play, it does not do so as a more fundamental theory, something the unification view would expect. Biochemistry identifies mechanisms, but there is no question of subsuming cell biology under biochemistry in the way Kepler's laws were subsumed by Newton's. Piecemeal identification of diverse mechanisms for ineliminable biological processes is a far cry from the unificationist picture of subsuming more phenomena under fewer premises.

So we have good reason to doubt that unification is necessary for explanation. The laws in the social sciences may mostly be lower-level generalizations compared to the basic laws of physics, but we have seen no reason to deny that lower-level causal generalizations can explain.

We can reinforce this conclusion by looking more explicitly at the positive picture of explanation lurking behind the above discussion. On the view advocated here, explanation in the social sciences and much of the natural sciences results from identifying causes. There are many advantages to analyzing explanation in terms of causation. Thinking of explanation as citing causes avoids well-known problems facing other pictures of explanation. For example, one version of the unificationist approach – the "nomological-deductive" model – equates explanation with subsuming the events under laws. Without further restrictions, this account allows irrelevant facts to explain. For example, the law "falling barometers are followed by a change in the weather" does not explain changes in the weather, even if they can be deduced from it. Demanding that we pick out real causal factors avoids that problem and others.[18] Moreover, causality seems an objective fact about the world, not us, in a way that "expectedness" or "unity" seems not to be. We also understand causality relatively well. If we do not know how to give a reductive philosophical account of causation in non-causal terms or if there are puzzles about sorting out complex causes, we nonetheless have an understanding of causation that we do not have of unification or expectedness. In practice, we have innumerable methods for sorting out causes. And there is little doubt that much science aims primarily to find causes; it takes a long and contentious philosophical story to picture science otherwise.

Causal explanation need not be solely or primarily about *mechanisms*. As we will see in detail in Chapter 5, explanation and confirmation can frequently proceed without citing underlying mechanisms. So to emphasize causal explanation is not necessarily to emphasize microlevel detail. Usually good causal explanation will in fact operate at multiple levels simultaneously in an integrated pattern. That in fact is just what we saw in Paige, for his large-scale macrosociological claims were backed up with lower-level mechanisms. And I shall also argue in Chapter 5 that many macrosociological causal claims cannot be eliminated in favor of some individual level accounts.

The causal picture does have a place for generality, albeit a subsidiary one. Generality comes from abstracting from real causal complexes. Those generalizations will usually hold only *ceteris paribus;* they explain when

---

[18] For a survey of the problems with the covering law account of explanation, see van Fraassen (1981, pp. 103–130).

they describe real tendencies or partial causes. Some domains will allow for greater generality because common causes are present. Yet in the end the real explanatory work results from picking out the particular causes at work. Generalizations help in that process and result from it, but they are really derivative of the specific causal facts.[19] Thus though I used causal laws as a wedge into this chapter, the key factor in explanation, I would suggest, is not the laws but the causes. A well-confirmed account composed of entirely unique causes explains despite its specificity. Laws with wide scope are likely to pick out only very partial causes. They will be confirmed and will explain only to the extent that we are sure they apply – and that is done best by filling in their *ceteris paribus* clauses, frequently on a case-by-case basis. Thus laws or generalizations play a role, but a derivative one.[20]

Unification is similarly a secondary factor on the causal account. As we saw above, unification can be important as an evidential factor, as a heuristic in looking for causes, and as a by-product, assuming in the last two cases that common causes are likely. However, unification also has another role on the causal picture. We frequently can provide causal explanations only after we have reduced a mass of diverse phenomena to some kind of order. Finding a classification scheme that organizes motley observations sets the stage for pursuing explanations. Pre-Darwinian work on species classification, early work categorizing chemical elements, and Kepler's laws ordering planetary motions were all essential prolegomena to the explanations that followed. On the causal view, a perspicuous set of categories or a simple set of behavioral laws do not explain; they do not tell us why the phenomena happen, only what the phenomena are. But those classifications and regularities do show us what needs to be explained and thus are an important first step in identifying causes.

The substantive details of causal explanation will be largely domain specific. Explanatory power is, to use the terminology of Chapter 1, an abstract virtue, one that gets its real content through the empirical knowledge about causal processes in the relevant domain.[21] That knowledge, along with the kinds of pragmatic factors discussed in Chapter 2, will turn

---

[19] My claim here is not incompatible with the claim that every singular causal claim entails or even presupposes some law, so long as one recognizes that the corresponding "law" may not be at all general. If we drop any sharp epistemological distinction between laws and generalizations, then the question whether singular causes explain or are confirmed without laws loses much of its significance.

[20] See Little (1989) for other thoughts along these lines.

[21] My claim that accounts of explanation are substantive empirical claims draws in part on W. Salmon (1984).

the demand for causes into a much more specific picture of what good explanation involves.[22] So to a certain extent, any philosophical account of causal explanation is bound to be uninformative.

Of course this sketch is nothing like a full philosophical account, though others have made strides in that direction.[23] I have only given some quick reasons for thinking that unification approaches fail and that causal accounts succeed. Moreover, it may well be that our common sense concept of explanation is inherently ambiguous. Thus in later chapters I will pursue the strategy outlined in the last chapter: argue that the social sciences can provide explanations that either unify or reveal causes. However, my focus remains the prospect for good causal explanations in the social sciences. I think the causal account best fits the good work in the social sciences and for that reason gains added credibility.

Let us turn finally to the second concern mentioned at the beginning of this section – that most social science, even if it gives us causal generalizations, produces no real *theories*. Recall that Chapter 2 described three kinds of virtues a scientific account might have: evidential, explanatory, and formal. However, the view of social science I have been defending seriously downplays formal virtues. Though Paige's account might approximate a full-fledged "theory," much work in the social sciences does not; that work is far from a deductively organized system with universal scope. Rather, the social sciences frequently produce batches of particular causal explanations that bear little resemblance to what the positivists meant by a theory. Is this a fatal flaw?

I think not, for at least the following reasons:

(1) Theories – understood as formalizable systems of some complexity – have a relatively minimal role in at least some good natural science. Important parts of biology, for example, have nothing like a theory in the classical sense. This holds for less developed areas like ecology, but it also holds for cell biology as well. Both areas make many causal claims, but those claims are piecemeal *ceteris paribus* generalizations with no extensive deductive structure.

(2) The "theory" of a given research tradition is notoriously hard to identify and in fact most traditions have no single theory at all.

---

[22] This view can be further reinforced by considerations I have neglected here – namely, that science is a practice and practices typically rely on background skills, implicit knowledge, and the like, which are local in nature. For a suggestive discussion of these motivations, see Rouse (1987).

[23] Among them W. Salmon (1984) and Cartwright (1984).

Instead, scientists use a variety of roughly related models, depending on their goals and the context. For example, Kitcher's account of evolutionary biology both downplays "theory" in favor of argument strategies and recognizes numerous versions of the Darwinian approach. However, if there is sometimes no identifying "the" theory, then the positivists' ideal of a comprehensive, unified theoretical system looks like an imposition on science, not a reading of it. This is a Kuhnian moral pointed to in Section 2.3: if anything like paradigms are the guiding unit in science, they are both more and less than fully formalizable deductive systems.

(3) Confirmation and explanation can proceed without theories. Obviously even singular causal claims can be confirmed without us having access to any very elaborate theory, for we do so constantly in everyday life when we see that the rock broke the window or the nail punctured the tire. Moreover, the low-level generalizations that make up much social science do have numerous epistemic ties to other generalizations, specific facts, and so on. Such ties fall short of a deductively closed, axiomatized system, but they nonetheless provide room for cross tests, fair tests, and independent tests – as indicated by the fact that much testing in natural science depends on skills that may not be explicitly described or describable, on piecemeal knowledge, and the like, as Kuhn pointed out. Finally, we saw above that universality and unification are not essential to explanation, hence one main motivation for demanding highly developed theories is misplaced. So the causal accounts produced by the social sciences can provide well-confirmed *explanations* without providing extensive theories.

(4) "Theories" are sometimes really highly abstract models or, more accurately, families of such models, whose role is largely heuristic and pragmatic, not explanatory or directly evidential. Models with broad scope frequently require severe idealizations and are highly unrealistic as descriptions of actual phenomena. Population genetic models in evolutionary biology, exponential growth models in ecology, and the models of classical statistical mechanics are probably examples of unrealistic models whose role is more heuristic and pragmatic than explanatory. Those models allow researchers to think about, teach, and mathematicize highly complex phenomena.[24] When such models are used to explain, they

---

[24] On statistical mechanics, see Cartwright (1984, pp. 154–155); on population genetics, see Lewontin (1963).

must lose their broad and systematic character and become highly particularized, singular causal explanations (as Cartwright [1984] has argued). Thus, if abstract models serve heuristic and pragmatic functions, their absence in the social sciences is likewise a heuristic and pragmatic loss, not necessarily a sign of weak evidence and explanations.

For these reasons, good science can proceed without theories. Though the work of Paige and that of Hannan and Freeman (discussed in the next chapter) do approximate the positivist goal, there can be good social science without doing so, just as there is good biology that lacks elaborate formalized theories with tight deductive structures.

So we have now removed the last obstacles to this chapter's main conclusion – namely, that the social sciences can and sometimes do produce well-confirmed explanations. Earlier sections argued for that conclusion by removing conceptual objections, by showing that social science evidence can confirm, and by arguing that Paige's work handles its problems as well as good work in biology. This last section, by arguing that unification and theories can have a derivative role in good science, suggests that we can generalize our conclusions about Paige to other work in the social sciences. The next chapter further supports that suggestion by looking at social explanation in one of its most controversial guises – functionalism.

4

Functionalism defended

Chapter 3 defended naturalism primarily by defending *causal* explanation in the social sciences. Nonetheless, serious doubts remain about a science of society. Much social science does not trade in ordinary causal talk at all. Look, for example, at the great classical social theorists. Marx, Durkheim, Malinowski, and Parsons all relied heavily on teleological explanations – they explained social phenomena by citing their function or purpose, a function or purpose that usually no individual had in mind. However, the natural sciences became real sciences precisely when they gave up on such mystical explanations. Social scientists, on the other hand, have not dropped this pseudoscientific mode of explanation. Thus Chapter 3 defended only a small part of social science; the rest, a critic might claim, is still rotten to the core.

Answering these doubts is crucial for several reasons. First, functional or teleological explanations are part and parcel of social science as we know it and are likely to remain so. Nearly every tradition in the social sciences, from ecological anthropology to stratification theory to neoclassical economics, employs functional explanations. Moreover, such explanations are unlikely to go away. Much in the social world is the result of individuals pursuing their own interests, often while proclaiming some more selfless motivation. Much in the social world is the result of competition, both between individuals and between social institutions and practices. Much in the social world seems to have a life of its own; in short, to persist for reasons not obvious to common sense. Explaining such social phenomena leads quite naturally to invoking functions – to explaining the existence of social practices by the functions, explicit or hidden, that they serve. Elected officials behave as they do in order to promote their standing with interest groups. Small-scale societies living in harsh environments compete with other groups and with nature; their practices must exist in order to promote survival. The basic income distribution in the United States has not changed significantly in 100 years, so it must serve some purpose or function. Explanations like these – so common in the social sciences – are the natural outcome of trying to make sense of a complex social world driven by interests, purposes, and competition. Only a significant transformation in current social science is likely to eliminate them. Since my concern is to defend the prospects for science in the social sciences *as we now know them,* defending functional explanation is essential.

Defending functionalism is also important for two other reasons. Functionalist social science appears to be social science at its worst. Thus naturalism gains added support if social science in its most suspicious guise can meet standards of good science. Furthermore, successful functional explanations also support the holist cause. Functional explanations often proceed at the social level. They often explain why one institution exists by its role in or effects on the social world. As a result, functional explanations are often paradigm instances of macrosociological explanation. In fact, the functionalist work defended most strongly in this chapter – the work of Hannan and Freeman on the ecology of organizations – will provide a real-life counterexample to reductionist strategies in Chapter 5. So defending functionalist social science throws the gauntlet at antinaturalist critics and ultimately at individualists as well.

Aside from defending big theses, this chapter pursues another aim of this book – it ties philosophy of social science to concrete, ongoing controversies in social science research. I look at numerous pieces of social research, at the kinds of evidence that functional explanations may have, at weak and strong criticisms raised by social scientists, at the various different theses social scientists sometimes run together, and so on. These discussions hopefully again show that philosophy of the social sciences can and must be done in close contact with real social scientific practice.

As in previous chapters, my argument shall be that functionalist social science can both in principle and in practice be good science. The argument proceeds as follows. Section 4.1 outlines numerous criticisms: that functionalist social science does not provide mechanisms, that it is unfalsifiable, that it relies on an illegitimate analogy to natural selection, and so on. These worries about functionalism are considerably less *a priori* – and thus considerably more interesting – than the conceptual objections to social laws rehearsed in the last chapter. Answering them will be possible only after we have a clear account of functionalism and its evidence. Sections 4.2 and 4.3 take up these tasks. Section 4.4 looks at actual functionalist research and its problems. I illustrate how functionalist accounts can fail, how they can be badly criticized, and how they can succeed. As in Chapter 3, my ultimate goal is to show that some research produces relatively well-confirmed causal explanations. Section 4.5 returns to the criticisms raised in the beginning and answers them.

4.1 Functionalism and its critics

Social scientists use the term "functionalism" in diverse ways. At times the term refers to a specific theoretical movement in sociology and anthropology – the movement identified with Malinowski and Parsons, among others. At other times, functionalism refers to a general kind or schema of explanation, not to a specific theory. I will sort out these differences with care below. For the moment we can stick with a simple definition. In its broadest terms, functionalism explains social phenomena by means of their functions. Less trivially, functional accounts identify specific causal effects of a practice or institution and then argue that the practice exists *in order to* promote those effects. Initiation rites are a traditional example: anthropologists have argued that initiation rites exist because they promote social cohesion (F. Young 1962). Taken to the extreme – by Parsons and Malinowski, for example – functionalism becomes a total account of society. Essential social needs are identified, and then more or less every social institution is explained as existing in order to promote those needs.

Anyone familiar with sociology and anthropology knows just how important functionalism has been and continues to be. Functionalism dominated mainstream sociological and anthropological theory for most of this century. Not only Malinowski and Parsons but also Radcliffe-Brown, Durkheim, and Merton were functionalists. Marx often offered functionalist explanations. Current social theory is likewise strongly influenced by functionalism. "Neo-functionalism" is a trendy form of abstract sociological theory (Alexander, ed., 1985). On a more empirical level, functionalism permeates much work in sociology and anthropology. Faia's *Dynamic Functionalism* (1986), which primarily discusses methods for testing certain kinds of causal models, recently won a major prize in sociology. Current work in anthropology on small-scale societies and their environments explains social practices in terms of their functions. Current Marxist theories explain various institutions in capitalist society by their functions.

Among the many functionalist claims made by the above individuals and others are that:

> Ceremonial customs exist in order to transmit emotional dispositions essential for societal survival (Radcliffe-Brown 1977).
> The division of labor exists in order to promote social solidarity (Durkheim 1933).

The state exists in order to promote the interests of the ruling class (Miliband 1969).

Educational systems exist in order to promote capitalist interests (Bowles and Gintis 1971).

Welfare exists in order to promote labor discipline (Piven and Cloward 1971).

Pig slaughter among the Maring exists in order to promote ecological balance (Rappaport 1984).

Hindu taboos on eating beef exist in order to promote survival (Harris 1977).

Inuit hunting group size exists in order to promote survival (Smith 1981).

Inequality exists in order to ensure that the most important social positions are filled by the most qualified (Davis and Moore 1945).

Specialist organizations exist in order to take advantage of variable environments (Hannan and Freeman 1989).

Long-term contracts exist in order to promote profitability (Williamson 1975).

Social scientists have advanced many more such explanations. Functionalism obviously is a fundamental explanatory strategy in social research.

Functionalism is nonetheless as controversial as it is prevalent. A first source of doubt is the kinds of evidence functionalists provide. Elster (1983) and others argue that most functionalist social science is unconfirmed because functionalists do not cite the mechanisms connecting beneficial practices with their persistence (Vayda 1987; Little 1989, p. 61). Others doubt that functionalist analyses are really falsifiable: identifying the benefits of a social practice is too easy, as is finding "functional prerequisites" (Hallpike 1986; Elster 1983). Similarly, some argue that picking out useful practices is essentially value laden, for it presupposes some notion of proper functioning (Turner and Maryanski 1979). Doubts have also been raised about what little statistical evidence functionalists do offer (Hallpike 1986). The needed correlations seem to be lacking. Finally, practicing anthropologists sometimes claim that functional accounts are unsupported because alternative, ordinary causal explanations are available.

Equally numerous are worries about explanation. Teleological or functional explanations seem to explain present practices by future occurrences, a mystical process indeed. Moreover, functionalists frequently do not cite the mechanism underlying their explanation. Elster (1983, 1989) thinks this deficiency fatal, for (he claims) there can be no explanation without one. Functional accounts also do not explain how institutions arose, only why they currently persist; they likewise do not tell us why one of many different and equally useful practices exist. According to the critics, both complaints show that functionalism really does not adequately explain.

Functionalism's use of biological analogies is equally worrisome. Functionalists frequently cast their models in evolutionary terms. However, it is not obvious that they are entitled to do so. Societies and social entities seemingly do not reproduce, so it is not clear how natural selection–like mechanisms could operate (Hallpike 1986, R. Young 1988). New cultural traits also do not originate randomly, as they allegedly must if social processes are to parallel evolutionary ones (Ellen 1982). Finally, cultural change does not show any evolutionary pattern (Hallpike 1986). So, functionalists apparently borrow illegitimately from the theory of evolution.

If reasonable, these criticisms would tell strongly against functionalism in the social sciences and raise serious doubts about the prospects for a science of society. In the course of this chapter I try to answer these diverse objections.

4.2 What is functionalism?

Functionalism is a notoriously slippery doctrine. Yet we need a clear account of this doctrine to assess it. The "account" that follows is not, however, a traditional philosophical one. Traditional philosophical accounts seek a unitary conceptual analysis – one that provides the individually necessary and jointly sufficient conditions for all functional ascriptions in biology, the social sciences, and ordinary language more generally. Such an account is both unnecessary for our purposes and misguided. It is misguided because functional explanations may not be a unitary phenomenon across disciplines, because few human concepts seem to be defined by necessary and sufficient conditions, and because testing definitions against philosophers' linguistic intuitions does little to illuminate scientific practice. It is unnecessary since my aim is only to defend *some* controversial explanations *in the social sciences* that are commonly labelled functionalist. I do not need a unitary conceptual analysis to pursue that task.

It is helpful at the outset to distinguish between functionalism as a specific social theory and functionalism as a form of explanation. Functional explanations involve two broad claims: (1) that a social practice or institution has some characteristic effect and (2) that it exists in order to promote that effect. Functionalism as a theory à la Parsons or Radcliffe-Brown claims that most or all institutions exist in order to maintain social equilibrium or societal survival. Functionalism as a specific substantive theory is thus just a special case of functional explanation. Though there are many important and interesting issues about functionalist theory, I focus on functional explanation in general. Social scientists who have no truck with Parsonsian theory or its kind nonetheless often offer functional explanations; moreover, any evaluation of traditional functionalist theory awaits an understanding of functional explanation.

To help develop a model of functional explanation, I want to look first at four common accounts: those analyzing functions in terms of (1) roles or capacities, (2) circular causation, (3) homeostatic systems, and (4) selection by consequences. While there are social science explanations that conform to each of these ideas, none fully captures functional explanations in their most controversial form. They will, however, help direct us towards that end.

For early functionalist theorists like Radcliffe-Brown, explaining by appeal to functions meant identifying the role institutions and practices play in the larger society. Understood this way, ascribing a function involves identifying systematic effects in an interconnected system. Some contemporary philosophers have also argued that to cite a function is to assert that an entity has certain persistent capacities in an overall system (Cummins 1975). No doubt some explanations falling under the functionalist rubric do precisely this. However, this version of functionalism is of the least interest for our purposes. While finding the systematic effects of institutions may be both important and difficult, it is a quite ordinary species of ordinary causal analysis. To describe how kinship systems help determine residency patterns, for example, is to simply to propose that A causes B, *ceteris paribus*, a kind of claim defended in the last chapter. However, the functional explanations that critics find most objectionable not only describe effects, they also claim that things exist *because of* or *in order to have* those effects. In short, the role account does not get at the controversial part of functionalism I wish to defend.

A second account of functionalism comes from Faia (1986). In a recent book much praised by sociologists, Faia uses the idea of circular causation to analyze functionalism. He argues that feedback relations constitute the core of functional explanations. Factor A exists in order to promote B when A's effects on B in turn result in a causal process influencing A. For example, social contacts outside one's group lower prejudice; reduced prejudice increases contacts. Faia concludes that such explanations are widespread in the social sciences and thus that many social scientists are (frequently unwittingly) functionalists.

No doubt some social explanations rely on mutual causation; much statistical work – on what are called "non-recursive" causal models – has gone into finding tools to test such explanations. Confirming claims about

Functionalism defended 107

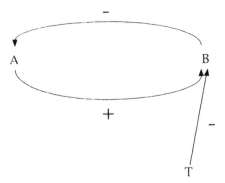

Figure 6: Homeostatic model of functional explanations. *A* exists in order to keep *B* in a given range, because when a threat (*T*) causes *B* to decline, *A* increases to re-establish *B* at the previous value.

mutual causation is certainly harder than confirming simple one-way causal connections. Yet in principle the defense of causal laws given in the last chapter should also extend to these more complex causal claims.

However, Faia's account does not capture those sorts of functional explanations we are after. Many ordinary processes exhibit feedback relations and yet clearly do not require or allow functional explanations. Feedback, as Faia describes it, ultimately is just circular causation. Given that every physical entity in the universe is causally interconnected with every other, numerous ordinary physical processes will be amenable to functionalist explanation on Faia's account. Surely Faia has cast the net too broadly for our purposes. Faia's analysis also is troublesome because it makes functional explanations "invertible." If *A* and *B* mutually interact, then their relation is symmetrical. So Faia's picture entails that if *A* exists in order to *B*, then it is equally true that *B* exists in order to *A*. That is an unhappy consequence as well.

A third approach to functional explanations focuses on homeostatic systems (Stinchcombe 1968; Nagel 1979). Imagine that the function of the liver is to control blood sugar levels. The homeostatic analysis says that livers increase their functioning when blood sugar levels are too high, decrease their functioning when blood sugar levels are too low. Put more generally, an entity *A* has a function *B* if whenever some outside factor *T* interferes to prevent *B*, then *A* would change so that *B* is re-established. Once the outside interference is removed *A* returns to its previous value. To have a function is to play this role in a homeostatic system. This homeostatic process is depicted in Figure 6.

Certainly some functionalist explanations in the social sciences fit this mold. Some Marxist theories, for example, describe a homeostatic system when they claim that particular phenomena exist in order to promote capitalist interests. Piven and Cloward's (1971) work on welfare and dissent argues that welfare spending exists in order to control dissent and promote labor discipline. They cash out this thesis by arguing that welfare spending increases in response to rising protest and returns to previous levels when the threat is removed. Marxists sometimes give similar accounts of unemployment and labor discipline, or military spending and economic downturns.

While the homeostatic account may capture some versions of functionalism, I shall not discuss it here in detail for two reasons. First, the homeostatic version of functionalism is not the most controversial variant – and my concern here is to defend naturalism at its allegedly weakest spot. Homeostatic systems are causal systems that are fairly well understood in principle. The system depicted in Figure 6 has a natural causal model:

$$T \uparrow \longrightarrow B \downarrow \longrightarrow A \uparrow \longrightarrow B \uparrow$$

Increases in the threat variable, T, cause a decline in B, which in turn causes an increase in A until T is removed. So pictured, the homeostatic process involves a straightforward series of causal processes, and we do not even need to invoke mutual interaction. Of course in practice confirming such models will confront all the difficulties facing causal models in general. Still, these problems are of the sort discussed in Chapter 3 and not ones especially associated with functionalism. While homeostatic processes deserve further investigation, other kinds of functional explanations need even more attention.

As the above remarks indicate, I also think that homeostatic processes are not good models for all or even the most prevalent forms of functionalist explanation.[1] The homeostatic account clearly fails for several key cases outside the social sciences. Some organ systems, for example the skeleton, do not change in the face of interfering outside influences. More drastically, traits produced by natural selection cannot be handled easily either. Larger finch beaks exist in order to ensure adequate nutrition. Nonetheless, if an outside force intervenes to change the food sources, we (and the finches) have no guarantee that finch beaks will change accordingly. Since many functional explanations in the social sciences are based on natural selection analogies, this problem is worrisome. Moreover, the homeostatic account does not fit actual social science practice. Functionalist work on group size, birth spacing, food sharing, and hunting strategies of small-

---

[1] For further arguments for this claim, see van Parijs (1981, p. 38).

Functionalism defended

scale societies does not employ the homeostatic metaphor but some sort of selectionist account. As we will see later, the same is true of good work in organizational ecology. And traditional functionalism à la Radcliffe-Brown and Parsons alleges that institutions exist because of their contribution to societal survival. In none of these cases is there any appeal to the homeostatic process sketched in Figure 6.[2]

The core metaphor behind functionalism holds that social practices exist *in order to* bring about their effects. To avoid future events causing present ones, functionalism is committed to some process tying an entity's causal consequences to its existence. In short, functional claims seem to presuppose some sort of selection by consequences. Cohen (1978) offers an interesting and well-known interpretation of functionalism along these lines. According to Cohen, functional explanations are a subspecies of consequence explanations – explanations that explain causes by their consequences. To establish that a social practice A exists in order to do B, we must establish a law that relates A's disposition to do B with A's existence. In short, we must show that it is a law that when A would be useful (or serve its function), A comes to exist. A functional explanation then is an explanation that invokes a particular sort of consequence law.

Cohen's account is helpful. It focuses more directly than previous accounts on the idea that a practice exists because of its function. Nonetheless, I think Cohen's analysis is inadequate as it stands. It is possible to have a well-confirmed consequence law (correlating A's function and its existence) without having an adequate functional explanation. The problem is simply that such correlations do not establish causation. To show that social practices exist when they serve useful functions does not prove that they exist because of those benefits. Until we rule out third factors, correlations between effects and existence are not sufficient.

---

[2] Those who think the homeostatic account is *the* right account might try to rephrase functionalist explanations that rely on selection by consequences in terms of a homeostatic process, taking survival as the goal and changes in the environment as threats. New traits would then be homeostatic responses to increase fitness. As suggested above, there is no real reason to think that natural selection must work this way. Threats may lower fitness without some new trait arising that re-establishes fitness levels – as extinct species illustrate. Also, there is no obvious analogy for the B-to-A causal loop in natural selection – old genes do not reappear when an environment returns to some previous state. Finally, it is arguable that the homeostatic account can be seen as a special case of the consequence account developed next – increases in A away from equilibrium cause movements in B towards equilibrium and persist because they do so. Since I think the search for a unitary account of functional language is a holdover from the days of *a priori* conceptual analysis, I do not push this argument here.

Consequence laws may also ask functional explanations to do too much.[3] Organisms that are mobile probably are so because mobility contributed to fitness. Yet, there is no reason to believe that when mobility would be useful, it comes to exist as Cohen's account requires. Plants no doubt would be better off if they could change locations to catch the sun, but structural constraints make that impossible. Similarly, large firms may exist because economies of scale contributed to their survival. Yet that does not mean that whenever a firm would benefit from larger size, it automatically becomes larger – lack of resources, foresight, and so on may prevent that. So consequence laws are also too strong a requirement for functional explanation.[4]

Cohen's account fails, but something like it is plausible. Taking a clue from Wright's (1973, 1976) "consequence-etiological" account of functions,[5] three components of such explanations seem to be:

[3] A similar criticism has been raised independently by Dickman (1990). Dickman assumes that Cohen's account is the only interpretation of functionalism and thus rejects functionalist interpretations of Marxism for other approaches, especially that of Hellman (1979). While Hellman's careful work may be a better rendering of Marx's theory, it obviously does not follow that functionalism is generally inadequate unless Cohen's account is the best understanding of functionalism, which it is not.

[4] Perhaps Cohen's account can be read in a way that avoids these problems. If we took consequence laws as causal laws, then the problem of law-like but non-causal factors would be eliminated. Cohen, however, apparently does not make that restriction, perhaps because he is wedded to some sort of nomological-deductive account of explanation (though not Hemple's version) or a Humean account of causation. Concerning whether consequence laws are necessary, Cohen might try to avoid the problem of mobile plants by requiring *real* dispositions rather than simply true counterfactual statements. Maybe plants and their environment do not have the relevant disposition. Yet Cohen's other examples do not favor this reading. He says that to show that expansion of scale exists in order to promote economies of scale, we have to confirm the law that whenever expansion of scale would cause economies, it comes to exist. But expansion might fail to occur when it would be efficient simply because of limits to human knowledge and so on. Still, this does not prevent expansion from occurring in other circumstances because of its contribution to efficiency. So unless every counterinstance to Cohen's version entails that there was no real disposition at work, it seems that requiring the dispositional facts to be taken realistically will not solve the problem.

[5] The account I use here borrows from Wright's insightful work but differs in several ways. One difference is that Wright does not explicitly require condition (3) and does not spell out (2) explicitly enough to be sure it entails (3); that means his account may be too broad – if ponds cause rain and vice versa and both persist in part because of that effect, then on his account ponds will exist in order to bring about rain. In short, Wright's account is too broad unless condition (3) is made explicit. Another difference is that Wright uses "*A* exists" rather than "persists," a difference motivated by problems I discuss below.

(1) *A* causes *B*.
(2) *A* persists because it causes *B*.
(3) *A* is causally prior to *B*.

The first condition is an ordinary causal claim. The second condition does much of the work, though the basic idea is relatively straightforward. A given social practice has a certain effect. When it has that effect, there is some causal mechanism that ensures *A* continues to exist. When the practice stops having that effect, that mechanism stops operating. The second condition is thus a causal claim – we are not appealing to future effects or to effects it would have if it existed. In the simplest form, we can imagine that at one moment *A* causes *B* and then that fact causes *A*'s existence during the next moment.

The third condition says that the causal process can be *initiated* with *A*, not with *B*. If *A* persists in order to *B* and does so for some extended period, then conditions (1) and (2) will produce a causal chain of this sort:

$$At_1 \longrightarrow Bt_2 \longrightarrow At_3 \longrightarrow Bt_4$$

Condition (3) says this chain can start with *A* and cannot start with *B*. If *B* is introduced, *A* will not result. In other words, *B* causes *A*'s persistence only if *B* was brought about by *A*. For example, imagine that initiation rites exist in order to bring about social solidarity. Condition (3) says initiations have a functional explanation only if independently introducing solidarity will not bring about initiation rites and start a chain like that above. Initiation rites have to be causally prior in this sense. The motivation for this third condition will emerge below.[6] (Note that condition [3] does *not* say that *A* is the only way to bring about *B*. If other factors exist in order to bring about *B*, then those causal chains cannot begin with *B* either.)

Focusing on these conditions avoids problems seen earlier. Adaptive traits of organisms, I argued, do not necessarily exhibit homeostasis – we have no guarantee that the trait in question will change correctly in the face of disturbing factors. Conditions (1) and (2) have just this result. If something interferes and prevents *A* from causing *B*, nothing guarantees that *A* changes appropriately. And unlike Cohen's dispositional account, the problem of spurious correlations is avoided, because condition (2) is a

---

[6] On one reading of condition (2), it may in fact entail (3). If condition (2) is read as implying that *B* alone is not sufficient for *A*'s persistence – in short, if condition (2) requires that it is the causing of *B* by *A* that explains *A*'s persistence – then condition (3) follows from (2). However, "the causing of *B* by *A*" may seem a funny kind of thing to be a cause and thus condition (2) might best be taken as simply that *B* causes *A*'s persistence. If so, condition (3) is still essential.

causal one. Of course, in trying to establish that these conditions hold, spurious correlations are a threat, as they are in any causal testing. But, spurious correlations are not allowed by the analysis itself, as in Cohen's account.

These conditions also ensure that functional explanations are not invertible and are narrower than mutual causation. Conditions (1) and (2) by themselves ensure that if A exists in order to B, it does not necessarily follow that B exists in order to A. More strongly, condition (3) explicitly *rules out* inverting the explanation. Conditions (1) and (2) alone are also incompatible with some mutual causation – specifically, runaway negative feedback. If A causes B and B drives A out of existence, then obviously the second condition fails. Condition (3) goes further and rules out positive mutual reinforcement. If A and B are simply mutually dependent, then there would be no reason to pick A as existing in order to B rather than vice versa. Condition (3) thus helps avoid Faia's problem – interpreting functional explanation so broadly that it loses its interesting and controversial character.

Obviously this picture of functional explanation has ties to evolution by natural selection. Yet it is important to see that differential sorting with heritability satisfies these conditions, but it is not necessary. Acquired characteristics passed on through learning or mimicking will also do, as will conscious human intervention.[7] Complex combinations of intentional action, unintended consequences of intentional action, and differential survival of social practices might likewise make these conditions true. Moreover, the above conditions make no requirement that A's function is beneficial to someone or something. So selectionist models with heredity are not necessary. Natural selection is thus really just a special case of the more general process that these conditions pick out.

Some philosophers doubt that consequence-etiological accounts of functional explanation are adequate. Many of their doubts turn on assuming that any adequate account must provide universal necessary and sufficient conditions, a task I explicitly eschewed. However, it will be useful to look briefly at some of the more common doubts, if only to forestall confusion. Criticisms found in the literature and responses to them include:

(1) *"A exists in order to B" cannot mean that A has a certain causal history, because the idea of selection by consequences is a recent idea but functional talk is not.*[8] Response: The above account of

---

[7] Note also that in some cases, condition (2) might hold because it is falsely believed that A causes B. Such situations are possible and would call for the appropriate modification of the first condition. The basic idea of course is still the same.

[8] Raised by Borse (1984, p. 373): "The modern theory of evolution is of recent vintage; talk of functions had been going on for a long time before it appeared."

functional explanation is not trying to produce a dictionary entry or predict philosophers' verbal behavior. Insofar as people in the past offered functionalist explanations without any clear account of them and thus without showing that the three conditions held, they were probably offering inadequate explanations.

(2) *We ascribe functions that clearly could not have arisen through selection by consequences or a similar process* (Borse 1984). The nose, for example, functions to keep eyeglasses in place. Response: It is important to distinguish, as we did earlier, functional explanations as citing causal roles versus functional explanations as claims that something exists in order to bring about its effects. The nose is a structural cause of wearing eyeglasses – noses play that causal role! Yet that is different from claiming noses exist to have that effect, obviously a false claim and one not equivalent to identifying their causal role.

(3) *My organs have functions now but do not exist because of their effects but because of the facts of development* (Achinstein 1980; Borse 1984). Here we have to be clear about the kind of explanation we are calling for – whether we are explaining types or tokens, whether we want proximate causes or ultimate causes, whether we are making a claim about existence at a moment or persistence over time. If we are talking about kinds of traits or practices (hearts or initiation rites), then those kinds do persist because of their effects. Even tokens – my heart – exist indirectly because of effects, namely the effects of hearts in the previous generation. And of course my heart also persists because of its effects now – namely, keeping me and thus it alive. This illustrates why "persists" in condition (2) is generally more perspicuous than "exists," which easily leads to confusion if not properly interpreted.

(4) *Causal-etiological accounts of "A exists to B" do not make A necessary for B*. Response: Why should they? Most causal explanations do not cite fully necessary conditions; functional explanations are no exception. This is in fact a virtue of causal-etiological renderings of functional explanations – it treats them no differently from other explanations in this regard.[9]

---

[9] Note that the requirement that functional explanations cite "causally prior" factors is not the same as making those factors the only way to produce the effect. If initiation rites exist to promote solidarity, then condition (3) above says that introducing solidarity will not cause initiation rites to come to exist. But that claim does not rule out some other process producing solidarity and persisting because it does so.

(5) *Accounts based on causal history make having a function contingent on past events, but items that are functionally explained would have their functions even if they came to exist five minutes ago* (Bigelow and Pargetter 1987). Response: Functionalism as a causal role claim does not require a causal history, but to capture the idea that *A* exists *in order to B*, we need something other than a causal role claim, namely, the conditions I cite. If large finch beaks (as a type) just popped into existence, then they do not exist in order to eat large seeds, though they may play that role.

(6) *Appeals to functions to explain persistence are circular, because if "A functions to B" means "A persists because it causes B" then saying A persists because of its function is saying A persists because it persists* (Bigelow and Pargetter 1987). Response: This confusion results only if we collapse functions as causal roles into functions as existing in order to. If we see that these are two different kinds of claims, then the circularity disappears. "Initiation rites function to create social solidarity" can make two different explanatory claims – that initiation rites cause solidarity only or that they exist in order to do so. If we are identifying causal roles, then we are not making a claim about persistence when we identify the function of initiation rites. If we are saying why initiation rites exist, then we are. We may sometimes use the same language to make both kinds of explanations, but they are nonetheless different. This criticism turns on ignoring the difference.

The causal-etiological account thus can handle common worries. Whether it handles all interesting cases of functional explanation in all domains is another question. I have already pointed out some explanations going by the functionalist label that really require only mutual causation. However, the three conditions listed above will do for our purposes here. They give us a model of the most controversial aspect of functionalism – namely, that something exists in order to have its effects. Having given some account of functional explanation, I now turn to the much more interesting question of how functionalist claims are confirmed and whether there are successful ones in the social sciences.

4.3 Confirming functional explanations

So far I have argued that even the most controversial forms of functional explanation are really a species of causal explanation. Thus there is nothing incoherent about the logic of functionalism. Yet functionalism might be in principle coherent but in practice unbelievable. To de-

Functionalism defended

fend more than just the possibility of naturalism, we need to show that functional explanations can actually be tested and confirmed and, ideally, that some functional hypotheses meet the basic standards of scientific adequacy. This section thus discusses what would count as evidence for functionalist hypotheses as well as some of its difficulties. Later sections apply these results to specific social research.

Evidence for functionalist claims falls on a rough continuum between direct and indirect proof. Direct evidence tries to show that the conditions for a functional explanation hold one by one; indirect evidence shows a functional explanation is compatible with another set of facts. Let me start first with the more direct kinds of evidence.

Recall our basic analysis of functionalism. A exists in order to B if (1) A causes B, (2) A persists because it causes B, and (3) A is causally prior to B. The most direct evidence thus tries to show these conditions obtain one by one. The first condition is a simple causal claim, and we assess it using the same methods applied to causal claims throughout the social sciences. The other two conditions are harder. They require us to establish that when A causes B, then A continues to persist – and that it does so because it causes B. So we look for cases where A brings about B. We then ask if A tends to persist thereafter and try to ascertain why it does so. If we control for other possible causes and still find that A's persistence is correlated with its causing B, then we have good evidence for the functional claim.[10] A correlation between A causing B and A's persistence means that B alone is not correlated with A's persistence and thus is unlikely to initiate the causal chain on its own, thus providing evidence that condition (3) holds as well as condition (2). Generally, background information will make condition (3) fairly obvious – it seems unlikely, for example, that creating social solidarity will cause societies to adopt initiation rites.

Put verbally, this may sound relatively straightforward. In practice, of course, things are not so simple. Since the best quantitative evidence in the social sciences is usually statistical, what does this mean in statistical terms? That question is complex and difficult. A simple first approximation at an answer is that functionalist claims must confirm something like the following causal model:

[10] This gives us evidence for condition (3) because a correlation between A causing B and A's persistence means that B alone is not correlated with A's persistence and thus is unlikely to initiate the causal chain on its own. Generally, background information makes condition (3) fairly obvious – it is clear, for example, that creating social solidarity will not cause societies to adopt initiation rites. Thus the second condition is usually the one most in need of evidence.

$$B = x_1A + e_1 \qquad (1)$$
$$P = x_2E + e_2 \qquad (2)$$

where A is the trait or practice to be functionally explained, B is A's function or effect, P is a variable measuring A's persistence, E measures whether equation (1) held at some time prior to P, x is the regression coefficient, and e is an error term. E might be a dummy variable simply taking on a value of zero or one, depending on the result of testing equation (1); P might use hazard analysis or some other tool to estimate the probability of A existing after the initial state. For example, B might be a measure of social solidarity and A a measure of the frequency of initiation rites. P would measure the duration of initiation rites, with E registering whether initiation rites cause solidarity as determined by the first equation. Then if we have good evidence for these two equations, we have good evidence that A causes B and persists because it does so. Good evidence obviously requires that other factors be controlled for, and in practice these simple regressions would not be enough. No doubt this model is statistically crude. But it does provide a rough generic model that gives us some idea how quantitative statistical evidence might support functionalist hypotheses.

In principle at least, we could test this model directly. Anthropologists, for example, have compiled a huge data bank on small-scale societies in the Ethnographic Atlas and the Human Relations Area File. Imagine that this data set contained not just cross-sectional data on hundreds of societies but also time series data measuring the duration of various practices. Then we might be able to identify and estimate the above model directly. Unfortunately, the HRAF does not contain the appropriate time series information. Still, direct statistical tests of equations like these might well be possible in other contexts. As we shall see later, Hannan and Freeman's work in organizational ecology does just that.

Direct evidence would also come from showing differential sorting or survival by traits. If we can show that some trait A of a social entity S causally contributes to S's survival, then we have initial evidence that A exists because of its function. Using the generic model above, think of A as initiation rites and B as some measure of survival. Then showing a nonspurious correlation between A and B is evidence for equation (1) and thus for the hypothesis that initiation rites cause survival. Still, showing that A contributes to survival is not good enough. If we want to show that A exists *in order to*, then we also need to show that contributing to survival causes A to persist. Differential survival does not *entail* persistence. Initiation rites could cause differential survival and still not persist – because the surviving social entities might rapidly change traits. Or, initiation rites could cause survival and could persist, yet not persist because of their effect on survival but for some other reason. So we still need evidence for equation (2). In Darwinian natural selection that evidence comes from establishing heritability.[11] Analogues in the social sciences would need some equivalent mechanism linking contributions to survival with persistence.

We thus see in more detail what I argued earlier: differential sorting models are special cases of the more general model of functional explanation. Showing that a practice causes survival and that those practices are in some sense heritable specifies *one* process that can bring about the two functionalist equations discussed above. Yet they are not essential to doing so. Any process that allows A to persist because it causes B will do – regardless of whether the process involves entities that reproduce, compete, and inherit traits. I dwell on this point, because I shall argue later that it removes some common criticisms of functionalism.

Other, less direct kinds of evidence are also available. Biologists often test for natural selection by finding a correlation between a trait and an environment. Strong correlations suggest that the trait exists because of the environment. This evidence is a first step toward establishing that the trait in question has been selected. Similar evidence may be possible in the social realm as well. If swidden agriculture occurs only in specific kinds of environments, then we have reason to think that there is something about swidden techniques that allow them to persist while other practices do not. In short, trait-environment correlation suggests equations (1) and (2) might hold. Ecological anthropology makes most frequent use of these arguments, but if we construe "environment" broadly to include social factors, this indirect evidence can be useful elsewhere.

Another indirect form of evidence also parallels testing in evolutionary biology. Biologists find evidence that a trait is adaptive by doing design analyses – by showing that a given trait would best solve the environmental problems an organism faces. If the analysis and the actual trait correspond, then we have a prima facie case for thinking that the trait exists in order to solve the relevant problem.

Design analyses generally take the form of optimality arguments, arguments that go roughly like this:

(1) A trait of type A would maximize fitness, would be the best available solution to the identified environmental problem, or the like.

---

[11] Endler (1986) provides a nice discussion of how evolutionary biologists establish these two broad conditions as well as the difficulties besetting the evidence. Social scientists might profit from a careful reading of his discussion.

118   *Philosophical foundations of the social sciences*

(2) There is a selective process that would result in the optimal trait being established.
(3) The observed trait is of type *A*.
(4) Thus the observed trait exists in order to maximize fitness, solve the identified environmental problem, or the like.

The social sciences also invoke optimality arguments – economics and anthropology employ them explicitly, traditional functionalist theory does so at least implicitly.[12] Obviously the key premises are the first two. Ideally, the argument that some trait is optimal would be theory driven. Microeconomic theory, for example, generates predictions about optimal levels of production and the relative use of factors. If one is confident about the best possible alternative and has good evidence that some process favors the optimal, then design arguments are powerful evidence for functionalism. However, as we shall see, these are not trivial assumptions.

Design evidence can also come from "stability arguments." Stability arguments identify traits or practices that no process would undo once established. Rather than pick *the* optimal trait, a stability argument has only to pick some list of stable traits. On the other hand, stability arguments must show that no other behavior could dislodge any stable trait. Evolutionary biology has used stability arguments with some success to show traits persist because of their effects. Many applications of game theory in the social sciences – for example, evolutionary approaches in economics[13] – try to apply them as well. As with optimality arguments, stability arguments depend on strong assumptions.

Thus functionalism can be in principle tested. Its main claims can be coherently formulated and diverse evidence at least in principle adduced. Whether the "in principle" can be made an "in fact" is an open question. All the testing procedures discussed above are fraught with complications. Those complications sometimes may entirely undercut the tests I have cited. At other times, however, tests may appear to disconfirm what are really plausible hypotheses because important complexities are ignored. I thus want to outline some of the complexities before turning to actual social research.

Design analysis and optimality arguments are controversial in biology,[14] so we should expect them to be even more so in the social sciences. An optimality argument can fail in at least four ways: (1) we may not accurately identify all the possible traits or practices, (2) we may misestimate

[12] See Kincaid (1995) for a discussion of optimality arguments in economics.
[13] See, for example, Nelson and Winter (1982).
[14] For an overview of the problems, see Horan (1990) and Kitcher (1987).

Functionalism defended

which of the possible ones is the best, (3) we may be wrong that there is a selective process at work, and (4) the relevant selective process may not select the optimal. Our evidence for a functionalist claim is only as strong as our confidence that all these errors are ruled out. And it will not do to argue that a practice must exist to serve a purpose or function just because we have estimated the optimal and found it present. Other processes can bring about optimal practices. Until we rule that out, the optimality argument has not shown the trait or practice persists because of its function.

Stability arguments avoid some of these difficulties but bring other liabilities. A successful stability argument need not identify *the* optimal trait; in this sense stability analysis requires less. Yet stability arguments can fail to accurately identify *all* the relevant stable traits. Furthermore, they have the added difficulty of showing that a trait really is stable – that once established, no set of circumstances could lead to its demise. Though stability arguments do not make bothersome assumptions about optimality, showing that selection optimizes may sometimes be easier than proving that a trait is stable. In prisoner's dilemma situations, for example, tit-for-tat seems to outcompete any rival strategy. Yet if the initial population is composed of individuals playing some "nice" strategy, then tit-for-tat can be invaded by pure cooperation. Yet pure cooperation loses out to nasty strategies that defect. So quite contextual factors can affect what is stable and what is not, and that makes it harder to warrant a functionalist hypothesis.[15] Thus design evidence – both from optimality and stability arguments – borrows on detailed background information. We can easily see how that debt might be more than the social sciences (or biology for that matter) can sometimes pay.

Trait-environment correlations have an equivalent prognosis. By themselves, they do not rule out numerous, non-functional explanations. Ultimately, they are conclusive evidence only if we are certain that some selective process underlies the tie between trait and environment. Controlling for other, non-selective factors that might make the correlation spurious can help. Nonetheless, doing so is much easier to recommend than it is to realize in practice. Trait-environment correlations are seldom decisive in evolutionary biology, where the presumption of selective processes is much stronger. So we can expect a worse prognosis in the social realm.

So far we have discussed complexities that make good evidence for functionalist claims harder to come by. Not all complexities have this effect, however. For much direct evidence, ignoring complexities is just as likely to make functionalism look weak when it is not. Recall our elementary causal model for functional accounts. That model assumes the world is a

---

[15] I take this point from Kitcher (1987).

very simple causal place. Each equation invokes only one independent variable. However, functional processes can be much more complicated. Multiple independent variables may be at work. Functional processes may interact with ordinary causal ones. Testing simple models when more complex ones are called for can produce misleading results – frequently results falsely disconfirming the functionalist hypothesis.

The simple functionalist model – where A causes B and persists because it does so – might be wrong in at least the following ways:

(1) A may have multiple effects, one of which is functional and the other dysfunctional;
(2) A may have multiple effects, each of which is by itself only weakly functional;
(3) A may persist only in part because of its function;
(4) A may not be the sole cause of B;
(5) the function B may not always be brought about by A; and
(6) A may persist because of its function in some contexts and for non-functional reasons in others.

Each of these complexities calls for more complex models; each can lead us to miss functional explanations where they really are legitimate.

Figure 7 depicts these frequently ignored complications. Using our rough causal model, the A-to-B link graphs equation (1), $B = x_1 A + e_1$. The B-to-P link represents the causal relation between causing A and persisting, that is, equation (2), $P = x_2 E + e_2$, where P measures persistence and E measures the prior causal effect of A on B. The complexities listed above basically mean that testing these equations gives misleading results; each situation requires that we add other variables. In Figure 7 situations (b) and (c) result because functional effects may happen on multiple levels or in multiple ways. Some practice A might have useful effects at an organizational, class, or social level for example – or its effects at one level might counteract those at another. (b) depicts just this situation. When the true model is as in (b) but we test for the elementary model, we may find equation (2) statistically insignificant. C is screening off the effect on persistence. If the real world is as in (c), similar but less severe confusion results from assuming the simple model: equation (2) again may be statistically insignificant, because we are ignoring other causes of A's persistence. The second condition can also fail in situation (d). Here A persists because of its effects but does not persist solely for that reason. Once again, employing the elementary model may cause us to miss a real relation. Better tests, of course, require adding the complicating factors to equation (2).

Situations (e), (f), and (g) cause more serious problems when we assume

Functionalism defended

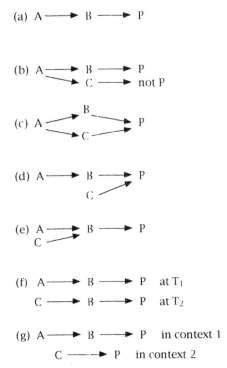

Figure 7: Various ways the simple model of functional explanation can be complicated. (a) The simple model, where A is to be functionally explained, B is its effect, and P is the persistence of A. (b) A exists in order to B but also has effect C, which is dysfunctional for A. (c) Multiple functions are present: A persists because of effect B and effect C. (d) A exists in order to B, but A also persists for reasons other than causing B. (e) A exists in order to B, but A is not the sole cause of effect B. (f) Functional equivalents: at one time, A exists in order to B, at another C exists in order to B. (g) A is functionally explained in one casual context but exists for non-functional reasons in another.

an elementary model. In (e), A is only partly responsible for its effect and using a simple model may lead us to reject equation (1) as statistically insignificant. It will also understate equation (2) for the same reasons as cases (b)–(d). Again, the other causal factors have to be added to the test.

Situation (f) depicts the problem of functional equivalents: at one time the useful effect results from A, at another from C. Testing for only the elementary model will result in both weak evidence that A causes B and that A's effect and its persistence occur at rates greater than chance. The elementary model also assumes that A has a uniform effect across all contexts. Situation (g) illustrates circumstances where that assumption fails: in some contexts A persists because of its useful effects, in others for other reasons. Lumping these two contexts together means that both equations may be weak. It will look like A is weakly correlated with B and that A's effect is weakly connected with its persistence.

Thus complications abound, both for indirect and direct evidence for functional claims. Yet recognizing these complexities also helps make the case for naturalism, because it shows in detail how functionalist claims could in principle be confirmed and that criticism of simple functionalist models need not compromise more sophisticated accounts. Moreover, the complexities identified above are in no way unique to the social sciences; they confront researchers in any discipline making complex causal claims. So the real question is once again not one of principle but practice: Can good social research handle these complexities? I turn now to such questions as we move from abstract models to actual functional explanations in the social sciences.

4.4 Functionalist failures and successes

Functionalist research runs the gamut from ad hoc speculation to carefully crafted, well-confirmed investigation. We survey some of this work below. I discuss both successes and failures, though only the former are strictly needed to defend naturalism. However, the failures are also worth investigating for two reasons. Seeing where functionalist research fails often shows how it could succeed and thus indirectly supports the prospects for naturalism. The failures are also interesting in their own right. The two failures exhibited here – optimality arguments in anthropology and the Marxian account of the state – are live controversies in the social sciences; applying the approach of this chapter to them should help clarify these ongoing debates. I begin with the worst failures, the optimality arguments, and save the best functionalist work, organizational ecology, for last.

4.4.1 *Optimal Eskimos and Hindu cows*

Functionalism has a long history in anthropology. Anthropologists typically work on small-scale societies or subsistence societies heavily dependent on the natural environment. As a result, design evidence has seemed quite appropriate – especially for explaining basic agricultural or hunting practices. Although short on details, many anthropologists assume that these practices persist through some sort of selective process. Thus the main evidence for their functionalist claims comes from showing that particular practices are ideally suited for the relevant environment. I want to discuss two functionalist accounts of this genre: Smith's optimal foraging models of Inuit hunting and Harris's functionalist explanation of Hindu beef aversions.

Anthropologists working on hunter-gatherer tribes have recently tried to import directly optimality arguments from biology. Smith (1981), a leading advocate of this approach, studied the Inuit Eskimos who live on the Canadian tundra. His goal, among others, was to explain Inuit hunting practices – in particular, the size of Inuit hunting groups. To this end Smith borrows models of optimal foraging developed in behavioral ecology. The models try to explain current behavior by deriving predictions from a model of patch type, food sources, energy expenditure, and so on. When the model correctly predicts observed behavior, biologists conclude that the trait probably exists because of natural selection. These are of course paradigm instances of optimality arguments for functional explanations. How do these arguments apply in the human case? Using net energy capture as the quantity to be maximized, Smith derives predictions about optimal hunting group size. By combining the basic optimal foraging model with what he learned about the details of Inuit hunting, Smith is able to generate several specific predictions. Comparisons with the data showed mixed results, with predictions holding for certain categories of hunt and faring less well in others. To the extent that these predictions bear up, Smith argues, we have reason to believe that hunting group size exists in order to promote energy capture.

This work is ingenious but unconvincing. A first problem is that Smith's predicted optimal sizes are doubtful because he leaves out so many factors. To name a few, Smith leaves out: (1) benefits resulting from hunts that have nothing to do with energy capture, for example benefits having to do with status within the group, with information and skill acquisition, with benefiting biological and non-biological kin, with solidifying relations leading to marriage, and with other benefits from social interactions; (2) factors affecting the importance of group size in energy capture, for example the relative skill level of individuals involved, the value of time spent in hunting versus other activities, the importance of average energy capture as opposed to reliable energy capture (high average capture might be less important than reliable capture of some prey), and so on; (3) costs besides energy expenditure, for example the probability of accident and injury; and (4) other complexities typical of natural selection, for example constraints placed on one practice by others, links between practices, possible group effects independent of individual energy gain, assumptions about equilibrium, and so on.

Aside from these questions, we also should worry about mechanisms. Smith asserts that causes can be found without citing mechanisms – a claim I argue for later. However, he does need evidence that there is *some* mechanism tying the caloric efficiency of group size with its persistence, for his optimality arguments depend on it. Yet we have no inkling what that might be. Smith builds his model as if individuals are maximizing individual genetic fitness and as if energy capture was a reasonable surrogate for contribution to fitness. Yet surely the mechanism is not genetic – can anyone seriously believe in a human gene that inclines Eskimos to perform different hunts in quite specific group sizes? Since Inuit hunters use not just energy in hunting but also economic resources, we would expect that any mechanism affecting group size would involve economic costs. But energy capture may have a tenuous correlation with the economic value of captured prey, so economic mechanisms do not seem to fit with Smith's model. Of course, it is possible that the Inuit have simply learned the optimal group sizes over time. But why would they learn energy-maximizing practices rather than those that minimize economic cost or, better yet, those that involve a combination of energy costs, energy expenditure, and other social benefits?

Thus we have doubts that Smith's model accurately predicts group size, ambiguous evidence about whether the predictions conform with the data, and no independent evidence that some maximizing mechanism exists. Given these weaknesses, we should have little confidence that group size exists to promote energy capture. Still, we can see how Smith might better confirm his functional explanation – by applying the general methods discussed in Chapter 3 for confirming *ceteris paribus* claims. We need to know that adding in more realistic factors leads to better predictions, that his predictions are stable when we vary the simplifications, that possible complicating factors are absent or have small influence, that there is good evidence for a mechanism tying functionality and persistence, and so on. Some efforts have been made in this direction (Smith 1985).

Marvin Harris (1979, 1985) argued for some different functionalist claims on roughly similar grounds. He advocates what he calls "cultural materialism" – the idea that we can fully explain human culture in economic, technological, and ecological terms. At his best, Harris takes apparently irrational and unfathomable social practices and shows their previously ignored economic or ecological benefits. One famous and apparently irrational practice is the Hindu beef aversion. Despite widespread hunger and starvation, Hindus refuse to eat beef. Harris denies that this aversion is irrational. Cows are not only a source of food. They are essential means for tilling land. For small plots, they are apparently more efficient than tractors. Cow dung is also a major fuel source. Finally, Hindu cow worship helps herds persist through droughts and famine, times when they otherwise might be decimated.

Harris gives us a functionalist explanation of beef aversion: beef aver-

Functionalism defended 125

sion exists in order to ensure the economic survival of Indian farmers.[16] His evidence for this claim is an optimality argument: cow worship maximizes agricultural success and survival prospects, therefore it exists because of those effects.

Optimality arguments succeed only if the practice in question is optimal and there exists some mechanism for selecting the optimal. Harris establishes neither claim. No doubt the beef taboo does have some clear agricultural and economic benefits. Yet demonstrating benefits is not enough. Harris needs to show that the total benefits outweigh the total costs. But he does not come close to establishing this stronger thesis. As his critics have pointed out, Harris ignores numerous costs of the beef ban (Vayda 1987). Wandering cows cause enormous crop damage in India. They also do not survive on thin air. Though Harris argues they do not compete much with humans for food, their diet significantly overlaps with that of the local population. Furthermore, supporting old, unreproductive cows has few of the benefits Harris describes. And supporting weak cows means that inferior cows are bred, another cost Harris ignores. So Harris's benefits may be overtaken by the many costs. In fact, the cattle population is lowest in the poorest regions, perhaps indicating that cow worship is a costly luxury.

Suppose for the moment that Harris's cost accounting were correct. We would still not know the cow ban was optimal. Or, more precisely, we could be sure the ban was optimal only if a complete ban and no ban were the only options. It is of course very hard to say what the possible religious taboos are. But is it obvious that less restrictive bans were impossible for India? What about bans that allowed some ritualistic killing of only very old cattle? Bans of this sort would cut out some of the costs and keep most of the benefits. I do not know if there was any precedent in Indian culture for other types of bans. Harris does not show that there was not, yet he needs to for a successful optimality argument.

Harris's problems do not stop here. Even if cow worship was optimal, we still need evidence that it persisted because it was optimal.[17] In short, we need to know there was some sort of mechanism tying benefits to persistence. Harris speculates that the beef aversion arose through cultural selection: families following the practice prospered and passed on their

---

[16] Harris is in fact not very clear on exactly what beef aversion maximizes – economic well-being, longevity, or some combination of the two.

[17] This point is made nicely by Vayda (1987). Though Vayda seems to think, along with Elster, that no functional claim is confirmed until we have the mechanism, his arguments are directed towards functional claims based on optimality evidence. Vayda's article helped me see that mechanisms can be essential when optimality evidence is involved, even if mechanisms are not generally required.

beliefs to their children, beef eaters did not. Such a mechanism is indeed possible and would do part of what Harris needs. Yet Harris has only weak and indirect evidence that such a mechanism existed. Alternative explanations see the taboo arising from above as an elite's way of promoting its interests.[18] From the outside, evidence for the latter account looks much stronger. Both accounts, however, are based on thin evidence. Thus even if Harris showed optimality, his case would still have some serious holes.

Finally, one well-known problem for Harris is that he largely ignores religion and other cultural factors. This complaint has at least two readings, one justified and one not. The justified concern is that cultural factors may explain why beef aversion persists rather than the economic benefits Harris describes. However, it would be unjustified to point to cultural influences and conclude that Harris's functionalist account *must* be wrong. While Harris may reject all cultural explanations, nothing about his functionalism per se requires it. As we saw in the last section, functional explanations can coexist with other causal factors. Thus the beef taboo might persist in part because of its effect on agriculture *and* for entirely nonfunctional reasons. So to point out the role of religious tradition is not necessarily a criticism, if we take Harris as citing a tendency or factor in a complex web of causes. Harris apparently does not understand his view this way. But he probably should, for it seems more likely to be true.

4.4.2 *Marxist accounts of the state*

Marx and subsequent Marxists explain the state in functionalist terms. In capitalist society, the dominant class is the bourgeoisie. They rule not only the means of production but society as a whole. In particular, the state – the military, the government bureaucracies, the legislative and judicial bodies – exists to protect the interests of the ruling capitalist class. This claim is a paradigmatic instance of functionalist explanation. Marxists are able to adduce some compelling evidence for this claim. Yet, those who picture the state differently severely criticize the Marxist account. I think we can show that neither side makes a very convincing case – largely because neither side has any very clear idea how functionalist explanations work.

Confusions beset this debate from the start. The disputed thesis – that the state exists in order to promote the interests of the ruling class – cries out for clarification. Neither "the state" nor "promoting the interests" is transparent on its own. The state is a complex entity that carries out innumerable activities. Surely Marxists do not claim that every state activity

---

[18] See Simoons (1979).

exists to promote capitalist interests. Many acts are simply neutral or unrelated to any clear class interests – or are traffic cops and septic tank inspectors really enforcers of labor discipline? So any serious Marxist claim – and any serious criticism of them – needs to specify types of state interventions at the very least. No doubt many other parameters need to be specified here as well – for example, whether all government actions are involved or simply the majority. Hence Marxists need a much more precise analysis of "the state." They also need to clarify "promoting interests." Again this idea involves many possible dimensions. Are we talking about the aggregate interest or the average or some other measure? Is "promoting" maximizing or is it meeting some baseline of "needs"? Social scientists must address these preliminary questions before any very serious empirical debate can be decisive. Although it is essential to point out these complexities, I shall put them aside. My concerns at this point are not about confusions implicit in "the state" or "promoting" but with the "in order to" that relates them.

Marxists support their view of the state with at least four kinds of evidence: (1) They argue that democracy in a capitalist system seldom if ever produces politicians willing to challenge capitalism. Elected officials may disagree over much, but they do not disagree over a fundamental commitment to preserving the capitalist system (Miliband 1969). (2) Marxists try to show that the modern state arose as part of the rise of the capitalist class. The American Revolution, for example, solidified the position of the capitalist class and in the process created a state suited to that purpose (Fisk 1989). (3) Marxists point to the various ways the state intervenes in society to preserve the capitalist system.[19] (4) Finally, Marxists identify the capitalist ties of those who staff state positions (Miliband 1969; Domhoff 1967). The third kind of evidence predominates, the fourth is usually the most empirically rigorous.

This evidence does not suffice. Showing that the state *arose* to promote capitalist interests does not show that it *currently* exists to perform that function. Origin and present function are two different things. Despite its capitalist heritage, the state might come to have different effects and to persist for different reasons. Marxists certainly want to allow this possibility when it comes to the *socialist* state; the Soviet state may have arisen to promote the interests of the working class, but only the most ideologically blinded think it had that function throughout its existence.

Furthermore, suppose Marxists are right that politicians do not chal-

[19] Nearly every Marxist account takes this evidence to be compelling proof that the state exists in order to protect capitalist interests. See, for example, Fisk (1989) and Miliband (1969).

lenge the capitalist order. While the Marxist account of the state might predict this behavior, other theories do so as well. If politicians pursue their own interest, how should we expect them to behave? Radical views of any stripe generally appeal only to a minority of the electorate. However, a politician's first priority is getting re-elected. So politicians may not be promoting capitalist interests but their own in the face of the facts about voter behavior. This explanation may of course be wrong – but it and others like it have to be ruled out. Pointing to "capitalist" politicians is not enough.

The capitalist ties of state officials are certainly better evidence. Yet it goes only so far. For example, those who see the state as an independent entity pursuing its own interests will not be convinced. After all, the ties are symmetrical – state officials have capitalist ties and capitalists have state ties. Thus it is possible for critics to argue that this evidence supports an opposite functionalist thesis: namely, that the capitalist economy exists to promote the interests of the state. In effect this evidence cites a correlation between A, state actions, and B, capitalist interests, and tries to argue that A exists in order to promote B. But the correlation, as we have seen, does not by itself tell us which way the causal arrow runs between A and B.

Furthermore, even if we can show that ties between the capitalists and the state benefit the former rather than the latter, we have done only half the job. If the state promotes the interests of capitalism, it does not follow without further evidence that it persists because it does so. Recall our previous simple causal model of functionalism. We need to establish both equations of that model; we need to show not just function but that performing a function causes persistence. Marxists, like the functionalist tradition generally, all too often give evidence for the first and conclude the latter.

It is worth pointing out for a moment why that gap must be closed. The problem is ultimately one of spurious correlation. The state might promote capitalist interests and persist without there being any causal connection between the two. Abstractly, consider the two causal systems involving the state, its function, and its persistence as diagrammed in Figure 8. A is the item to be functionally explained, B is its function, and P is A's persistence, with C being some other causal factor. In the first case, A causes B and its doing so is correlated with its persistence. Yet this correlation is spurious and A does not persist because it causes B. B and P are the result of a common cause, resulting in a spurious correlation. In the second case, A's causing B not only results in its persistence but in other effects as well; they too are thus spuriously correlated with A's persistence. In both instances, we can know that A has certain causal effects *and* persists when it does so. Yet we would be wrong to conclude that A persists *because of* these effects.

Functionalism defended

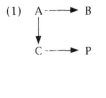

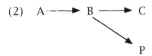

Figure 8: Two ways in which A's causing B can be spuriously correlated with A's persistence.

These situations are not just abstract possibilities. Other, non-Marxist explanations of the state would expect precisely such relationships to occur. For example, one standard non-Marxist theory claims that the state is an autonomous entity pursuing its own interests.[20] On this view, the state is not a mere tool of the dominant class. Nonetheless, the state may on occasion find it useful to promote the interests of the upper classes. Such circumstances result in precisely situation (1) depicted in Figure 8. The state pursues policies in order to promote its interests, but those policies have side effects that promote capitalist interests. For example, the state might play one group off another to maximize its interest. The fascist state is a particularly vivid instance of this phenomenon. Hitler broke the unions, but he did so, the story runs, not to promote capitalist interests but his own – and he was perfectly willing to abandon capitalist interests when they conflicted with those of the state. So situation (1) is a live possibility.

Some Marxists, including Marx, have seen that the state does not always act in the direct interests of the capitalist class. In particular, capitalist states sometimes seem to favor working-class interests – witness the New Deal and the Wagner Act, which set off the CIO organizing drives in the United States. The usual Marxist response calls for a more subtle analysis of the state. The state, the Marxist claims, defends capitalist interests by maintaining the capitalist system. Sometimes maintaining the system requires taking the opposite side. However, by preserving the basic system, capitalist interests are served. The New Deal and the recognition of labor unions did help channel labor militancy; the system was preserved and

---

[20] See, for example, Skocpol (1979) or more generally the work of the public choice school.

capitalists' long-term interests as a class were advanced. So evidence that the state preserves the basic social structure is evidence that it functions in order to promote ruling-class interests.

This evidence is also inconclusive. Assume that states which maintain social stability – and thus capitalist interests – are the states that persist.[21] Have we shown therefore that the state exists to protect capitalism? Situation (2) in Figure 8 shows that we have not. Again promoting capitalist interests could be a side effect and not the real reason the state exists. And again one competing explanation envisions precisely this circumstance. Chalmers Johnson's *Revolutionary Change* (1966) explains the state as existing to ensure social integration. This Parsonsian perspective might mean that the state promotes capitalist interests by promoting stability. Yet it is promoting stability, not capitalist interests, which allows the state to survive on this view. So Marxist accounts must rule out these alternative explanations. Until they do so, pointing out that the state benefits capitalism is not enough.

Critics of the Marxian account should take little satisfaction in this conclusion, however, for many attacks on the Marxian account face similar problems. As we have seen, Marxian explanations are standardly criticized for failing to see that the state has its own interests and for ignoring the obvious ways states promote non-capitalist interests.[22] Neither point need be a fatal objection to *subtle* Marxist accounts. Only the simplest functional model requires that the state exists solely to defend capitalist interests. More complex models allow functional relations to promote opposite tendencies, allow non-functional variables to play a role, allow functional explanations to hold in some contexts and not others, and so on (recall Figure 7). A sophisticated Marxist account could incorporate such complexities, perhaps with the stipulation that promoting ruling-class interests was the most important factor in some sense.[23] Nothing in functional explanation precludes other factors from working – any more than natural selection precludes genetic drift from operating in the biological world. Functionalist claims, like all causal claims in the social sciences, are tendency claims. Thus pointing out the non-functional aspects of the state is not a conclusive refutation.

So what does this debate show about functionalism? It does not show

---

[21] Earlier I noted that "promoting capitalist interests" has multiple possible meanings. That ambiguity shows up here. If the state exists only to preserve the capitalist system, then on this theory the state need not *maximize* capitalist interests at all, only favor them to the extent necessary to keep the system going.
[22] The first point is found in Skocpol (1979) and the second in de Jasay (1985, p. 55).
[23] See Hellman (1979) for an attempt to explicate a meaningful sense of "most important."

Functionalism defended

that functionalism, including Marxian functionalism, is in principle flawed. The flaws are in the practice. No doubt successfully evaluating Marxian functionalism is hard, both conceptually and empirically. Yet the difficulties discussed above likewise tell us what further evidence is needed. Marxists and their critics have often not taken those further steps, but that is not because there is no work to be done.

4.4.3 The ecology of organizations

So far I have surveyed functionalist failures, arguing that their failings are contingent and eliminable. In this section I argue for a much more ambitious claim – namely, that some functional explanations in the social sciences are good science. My evidence comes from a branch of sociology known as "organizational ecology," which, as the name implies, uses ecological models to explain organizational change. While the organizational ecology literature is vast and expanding (Carroll 1987), I shall focus in detail on the work of Hannan and Freeman (1989). Their research is arguably the most developed work in organizational ecology, both theoretically and empirically.

Organizational ecology seeks to explain the kinds of organizations that exist, their relative numbers, and how those kinds and relative numbers change. In short, organizational ecologists are interested in *populations* of organizations and the factors influencing their composition. Such population-level analysis is necessary, Hannan and Freeman claim, if sociologists are to explain large-scale social change.

Hannan and Freeman doubt that past work on organizations can explain large-scale trends. Earlier work focused largely on individual organizations at a given time, looking for traits typical of successful organizations and then extrapolating to the overall population. Yet this approach rests on numerous suspicious presuppositions, according to Hannan and Freeman. First, studying successful organizations easily leads to selection bias. The traits characteristic of successful organizations may or may not be the cause of that success. To make inferences about causes, we need to know what unsuccessful organizations were like – and thus we need to study population processes from the beginning. A second, questionable assumption needed for such inferences is that an equilibrium has been reached. If the population of organizations is undergoing a change, then the traits that are widespread at a given moment may not be there because they contribute to success – they may be on their way out. So cross-sectional data, which are data at a given instant in time, may be misleading when it comes to inferring the causes of population diversity. Thus explaining the kinds and numbers of organizations requires studying populations over time.

Most important of all, past work assumed that organizations could rapidly react to changing environments. Organizations were treated anthropomorphically – as "recognizing" problems imposed by the environment, identifying strategies, and then picking and acting on the best. Hannan and Freeman think this picture drastically overestimates organizational flexibility. Inertia, they believe, is generally the rule. Sunk costs, restraints on the internal flow of information, internal politics, individuals pursuing their own good at the expense of the organization as a whole, norms of behavior based on past history, legal barriers, and many other factors make flexible, rapidly changing organizations unlikely. So the processes determining organization success must lie elsewhere.

Hannan and Freeman propose an alternative picture inspired by models in evolutionary biology and ecology. Organizations compete for resources – for financial support, for legitimacy, for employees, and the like. If that competition is severe enough, then not all organizations can survive; individuals will likewise be discouraged from founding new ones. If organizations are slow to change in the face of environmental threats, then differential founding and survival will determine the kinds and numbers of organizations. In short, the social environment will lead to the survival of the fittest.

If a differential sorting mechanism lies behind organizational diversity, then models from evolutionary biology and ecology may suggest useful hypotheses about social organizations. For one, the notions of a "carrying capacity" and "density dependence" should apply. A particular social environment will have an upper limit to the number of organizations it can support. Competition will be greater as that carrying capacity is reached – survival chances will depend on the total size or "density" of the population. Survival will also depend on the density of other populations if different kinds of organizations compete for similar resources. So these familiar notions from biology and the generalizations connecting them should find a parallel in the social realm.

More specifically, ecological models describing niches and the strategies organisms use to exploit them should find parallels in the organizational world. A niche, in its simplest terms, is a set of resources. If that set includes a variety of different resources, the niche is wide; if the set involves only very specific factors, the niche is narrow. Organisms relying on wide niches are called generalists; those relying on narrow niches, specialists. Hannan and Freeman apply these concepts and the generalizations they suggest to organizations. Some organizations depend on a wide array of resources and thus are, in ecological terminology, "generalists." "Specialists" are those organizations that rely on quite specific resources.

To derive specific hypotheses about organizations and their environ-

Functionalism defended 133

	Stable Environments	Variable Environments	
		Coarse Grained	Fine Grained
Specialist Strategies	present	absent	present
Generalist Strategies	absent	present	absent

Figure 9: Hannan and Freeman's hypotheses about the effects of stable and variable environments on organizational strategies.

ments, Hannan and Freeman borrow from some work on patchy environments by Levins (1968). Levins argues that niches include not just kinds of resources but also their variability. Resources may be provided on a steady basis or they may fluctuate. Environments with steadily provided resources are called stable environments; environments with fluctuating resources are called variable environments. Variable environments themselves are not all alike, for variation may come in large "lumps" or may be evenly dispersed. Evenly dispersed variation produces "fine-grained" environments while variation that comes in patches results in "coarse-grained" environments. These different kinds of niches should produce different kinds of strategies. Specialist organizations will do well when environments are stable, for then the gains from specialization will outweigh the potential advantages of flexibility that accrue to generalists. However, generalist organizations are at an advantage in variable environments, unless the variation is fine grained and thus predictable; specialists will be able to take advantage of this kind of "stability." This analysis leads to specific hypotheses about when and where we should see generalist and specialist strategies. These hypotheses are summarized in Figure 9.

Hannan and Freeman's work thus produces functional explanations. Their explanations are functional since they assert that organizational traits exist in order to bring about their effects. Generalist strategies exist in order to take advantage of coarse-grained environments; specialist strategies exist in order to exploit fine-grained and stable environments. Like Darwin, Hannan and Freeman are able to give these apparently mysterious claims an ordinary causal reading based on variation and differential survival, though without employing any very specific analogues of reproduction or heritability. The process they describe is an instance of the abstract schema for functional explanations outlined earlier.

Hannan and Freeman test their theory in two basic steps. They first look for evidence that survival and founding of organizations is density dependent. If the chances of survival or founding depend on the size of the existing population, then we have a first piece of evidence that selection is occurring. If the chances of survival or founding depend on the size of other populations, then we have evidence both for selection and for what is "doing" the selecting. So establishing density dependence is a useful first step towards showing their general theory relevant to organizations. The second step aims for much more compelling evidence, namely, evidence that particular characteristics of organizations persist because of their contribution to the survival and founding of organizations. Here Hannan and Freeman concentrate on their hypotheses about specialist and generalist strategies.

Hannan and Freeman's evidence for density dependence and thus indirectly for selection comes from separate studies on national labor unions, semiconductor firms, and newspapers. Data on unions cover all national labor unions in the United States from 1836 to 1985. All U.S. semiconductor firms from the beginning of the industry in the 1940s to the present constitute the second data set. Data on newspapers are more localized, covering seven urban areas from 1840 to 1975. These data sets measure date of founding, date of disbanding, the size or density of the population at each time during the period, the size of a competing population at each time during the period, and a variety of environmental variables, for example, changing economic conditions in the case of labor unions.

From these data, Hannan and Freeman can then look for evidence of competition and density effects. Hannan and Freeman propose that foundings will initially increase with population size and then decline. When a type of organization is new, there is competition for legitimacy, and growing numbers contribute positively to the prospects of founding and survival. However, as the population grows, competition for resources takes over and increased density depresses the prospects of both founding and survival.

All three data sets support Hannan and Freeman's hypotheses. Foundings and survival probabilities increase with small population size and are depressed by further growth. Competing forms – national unions versus craft unions and subsidiary firms versus independent semiconductor firms – also depress the prospects of founding and survival. Hannan and Freeman also take great care to be sure these results are not artifacts of other processes. They see whether these relationships hold when environmental changes are controlled for, on the thought that changes in economic conditions, for example, might be a confounding factor. They control for age effects – effects that may hold for any organization over its life history regardless of the effects of competition. They use different specifications of their statistical equations and different measures of their variables to see if their results are "fragile," that is, if they are an artifact of the way the test is set up. In all these checks Hannan and Freeman's results hold up. In short, their claims about competition pass cross tests, fair tests, and independent tests.

Hannan and Freeman next go on to test their specific hypotheses about selection processes. Those processes, recall, concern different kinds of environments – stable and variable, fine-grained and coarse – and the organizational strategies – specialist or generalist – they favor. Data for these hypotheses come from the previous study of semiconductor firms and from a separate study of restaurants. The latter was a three-year prospective study of 1,000 restaurants, selected randomly from eighteen metropolitan areas. Prior to collecting both data sets, Hannan and Freeman specified criteria for specialist and generalist organizations and for stable, variable, fine-grained, and coarse-grained environments. Specialist restaurants are those with a narrow price range, hours of business, or range of products. Specialist semiconductor firms have a narrow range of products. Hannan and Freeman use variation in sales to measure stable and variable environments. When the variation has long "runs," the environment is coarse-grained; when the runs are short, the environment is fine grained. For a restaurant, for example, seasonal variations in sales mean coarse-grained variability while variations lasting less than a month mean fine-grained variability.

Using these specifications, Hannan and Freeman test their specific hypotheses about the causes of organizational mortality. Like good evolutionary biology (cf. Endler 1986), these studies control for potentially confounding factors such as age effects, size effects, local variations in the environment, and measurement artifacts. The data strongly support their hypotheses. Specialist organizational strategies promote survival in stable and in fine-grained variable environments and persist because they do so. Generalist strategies promote survival in variable environments and persist for that reason.

Hannan and Freeman's work is thus an exemplary piece of social research, and it shows that good functionalist social science is possible in practice. No doubt questions can still be raised about their work. Yet any remaining questions, I would suggest, are the sorts of questions that will plague any non-experimental discipline – doubts about uncontrolled variables and the like. As we saw in Chapter 3, those questions equally confront paradigm work in evolutionary biology and ecology. Grant's best evidence, for example, never reaches the degree of rigor found in Hannan and Freeman's three-year prospective study of a 1,000-member random sample. Far from being pseudoscience, this functional explanation is arguably as well confirmed as good work in the non-experimental natural sciences. Those who claim that the social sciences are doomed to failure have a lot of explaining (away) to do.

4.5 The critics answered

I turn now to answer the traditional criticisms raised in Section 4.1, many of which we have already implicitly addressed. As I noted earlier, these criticisms roughly divide into those concerning evidence, explanation, and parallels to natural selection. I begin with doubts about the evidence.

Elster and others have claimed that no functional explanation is confirmed until we have cited the specific mechanism involved; however, much functional explanation in the social sciences provides no mechanisms. While we can share Elster's concern for rigor, his claim is nonetheless false. "The mechanism" is, of course, vague – vague in ways that immediately belie Elster's requirement. Is knowing that there exists a selective mechanism sufficient? Or must we know the precise process? And, if so, in how much detail? Do we need to trace each causal step from effects to persistence? At what level of description? Do we need the quantum mechanical details? Once we begin asking such questions, it is obvious that the demand for mechanisms has no natural stopping point. Certainly the more detail, the better – yet something short of the whole story has to be sufficient, or else we would never confirm any causal claim. In other words, the problem is this: Functional explanations are causal explanations, but we can confirm causal claims without citing underlying mechanisms. So functional explanations are not necessarily unconfirmed until the mechanism is cited.

We can in fact have fairly good evidence for a functional claim without knowing the precise mechanism. To know that A persists because of its effects, I need to establish a non-spurious correlation between A's effect and its persistence. Elster is right that such a correlation might be spurious and that identifying underlying mechanisms will help avoid that problem. He is wrong, however, that identifying mechanisms is the only way to eliminate spurious causation. We can eliminate spurious causation by controlling for potentially confounding factors and we can do that without knowing the precise mechanism, whatever that means. Of course, this argument turns on the assumption that we can confirm causal claims without knowing their mechanisms. I have suggested intuitively why this assumption is reasonable; Chapter 5 will argue for it in more detail.

Elster's claim is perhaps more reasonable for indirect evidence. Indirect evidence is weaker and relies more heavily on background assumptions.[24] Optimality arguments and trait-environment correlations are compelling only if they rule out alternative, non-functional explanations. So we do need to know something about mechanisms in this case, unlike the general case of confirming functionalist hypotheses. Nonetheless, we still do not need to know the precise mechanism, only that some mechanism or other is at work. So Elster's claim is wrong about functionalism in general, half right about optimality evidence.

A related complaint about functionalism is its reliance on prerequisite analysis. No doubt much actual functionalist explanation employing this notion fares badly. Evidence identifying functional prerequisites is really a kind of design argument. Functionalists like Parsons tried to deduce the necessary conditions for societal survival. When institutions were found that contributed to those necessary conditions, it was concluded that they persisted because of their contribution. As we have seen, these optimality arguments succeed only if they meet some stringent conditions, ones prerequisite analysis usually ignores. Usually no compelling case is given that the particular practice *best* promotes the need in question or that some mechanism exists causing the optimal to persist. So prerequisite analysis is troublesome – because it usually supplies weak optimality arguments.[25]

---

[24] Elster's demand for mechanisms originated in part in response to Cohen's account of functionalism. As we saw, Cohen's account allows in spurious correlations and thus Elster's criticism, *as a criticism of Cohen,* is more reasonable. Cohen also apparently fails to see that the demand for mechanisms can be reasonable in the case of optimality arguments, witness his remark that "If, for example, the pattern of educational provision in a society evolves in a manner suitable to its changing economy, then it is reasonable to assert that education changes as it does because the changes sustain economic evolution" (1978, p. 286). To show that educational changes promote economic change is not compelling evidence that they exist to do so unless we know that some mechanism exists tying persistence with useful effects. Given that there are many possible non-functionalist accounts for this possible connection, a functionalist mechanism cannot simply be assumed.

[25] Note that this is a different problem than that which is usually raised – that prerequisite analysis is empty or unsupported. Prerequisites of social survival can be confirmed by looking for correlates of survival. Maybe no such universal prerequisites exist, but if we specify the kind of society or institution involved, then prerequisite analysis could ground optimality arguments without being universal – just as optimal foraging theory in biology might identify different components of fitness for different species of habitats. Many alleged societal prerequisites are indeed "trivial" in that they are obvious and uninformative by themselves. However, being obvious is not the real problem. The real problem is that identified above – that prerequisite analyses are design arguments that give little or no evidence that the necessary background conditions hold.

However, if prerequisite analysis fails, there is no inherent reason why optimality arguments in general must do so. We saw in Section 4.4 that there are a clear set of assumptions that will suffice to make a design argument compelling. Admittedly, good optimality arguments are hard to come by – in the social sciences and out. Fortunately for functionalism, they do not exhaust the possible evidence.

Some social scientists have doubted whether the idea of beneficial effects can be objectively evaluated – doesn't it presuppose the idea of proper functioning? And doesn't the "proper functioning" of an institution presuppose some conception of what is a good institution? Perhaps this problem plagues some traditional functionalist theory. However, it need not. As defined here, functional explanation does not presuppose *beneficial* effects – only that A causes B and persists because it does so. Even when beneficial effects are cited, we can sometimes specify in objective terms what we mean by "beneficial." Frequently, beneficial effects are those that contribute to survival. For organization ecology, beneficial effects are those that promote organizational survival and foundings. In Marxist functionalism, the beneficial effects come from promoting capitalist interests. "Organizational survival" and "capitalist interests" are of course hard to operationalize. Though Marxists have not really grappled with this problem, defenders of organization ecology have. And yet they do not make normative assumptions in any obvious way that I can see.

Some critics believe functional accounts too easy – almost any social practice is bound to have some positive effects. Probably this claim is true and probably much functional explanation has been weak in this way. In the last section we saw how Marxist accounts floundered on this problem. Nonetheless, functionalism need not make such errors. Organizational ecology is a case in point. Hannan and Freeman do not simply look for some beneficial effect flowing from organizational structures and then blithely posit that they exist because of those effects. Instead, they use powerful statistical techniques to find traits that seem to cause differential survival, controlling for alternative explanations. So positive effects are measured and the tie to survival is quite directly demonstrated.

Similar complaints – that statistical evidence is weak or that alternative, non-functional causes can be cited – get similar answers. Statistical evidence is sometimes weak or even non-existent. Yet that is not inevitable, as the work of Hannan and Freeman indicates. And, as I argued above, weak correlations may sometimes mean only that overly simple models are being tested. Likewise, non-functional causes can be a problem, but not because they show all functional explanation otiose. Functional accounts do not exclude multiple causes, even where the other causes have nothing to do with selection by consequences. Of course, if alternative explanations are not ruled out or incorporated, a functional account has weak evidence. Then the problem is with the singer, not the song.

Is perhaps the burden of proof on the shoulders of the functionalists? If a non-functionalist explanation is possible, then isn't the functionalist account automatically defeated? Not usually. The mere fact that we can imagine a non-functionalist explanation is no more telling than the fact that we imagine a functionalist one. If either has any prior credibility, then it must be ruled out to prove the other. So simply pointing to a conceivable non-functionalist alternative does not refute the fucntionalist explanation. However, there is a grain of truth in this worry. When functionalists rely entirely on optimality arguments, then citing plausible non-functionalist explanations raises real problems. Optimality arguments assume a mechanism exists. Pointing out other possible kinds of causes means that the needed mechanism may not operate at all. So alternative non-functional explanation may raise doubts about design evidence. Yet since functional and non-functional causes can coexist, imagining possible non-functional causes simply points out that we need more evidence. Possible non-functional causes do not by themselves decisively refute a functionalist hypothesis.[26]

Is there something about functionalist accounts that does not satisfy strictures on explanations (beyond problems of evidence)? Functional explanations are, I have argued in detail, a species of causal explanation. So in form at least they are perfectly acceptable. Do we perhaps need to cite mechanisms before a functional account explains? Since we can search for mechanisms on indefinitely many levels, this demand is also too strong. If I have good evidence that some organizational trait causes differential survival of those organizations and that those organizations do not rapidly change traits, then I can say why that trait exists – even if I do not know precisely how it causes differential survival. Mechanisms are not an essential requirement for explanation. (We shall defend this general claim more rigorously in Chapter 5.)

Hallpike was worried that functional explanations are inadequate because they do not explain the origins of institutions. We can grant his complaint, properly understood. Darwin could not explain how new traits arose. To that extent his theory was incomplete. However, an incomplete theory is not necessarily a false one. In other words, when a theory cannot answer some questions it does not follow that it cannot satisfactorily answer any. Social scientists might be able to show that practices or institu-

---

[26] Homans and Schneider (1955) argue against Lévi-Strauss's functionalist account of cross-cousin marriage by postulating an alternative explanation and showing it consistent with the data. As they acknowledge, such evidence is indecisive.

tions exist because of their effects – and yet have no account of how those practices first originated. Failure in the latter case need not mean failure in the former.

Hallpike is also worried that functional accounts are not sufficiently specific. In many cases, other social practices could serve the same function. Thus it seems an appeal to functions does not say why this practice exists rather than some other equivalent institution. Hallpike's conclusion seems plausible, but I am not sure what it shows. We may not be able to explain why *A* rather than *B* originally arose. That does not mean, however, that we cannot explain why *A* persists rather than not. We saw in Chapter 3 that explanatory questions must be specified along a number of implicit parameters. Among those parameters are the contrast class. "Why did Adam eat the apple?" can be understood in various ways: why did he *eat* (rather than throw, etc.) the apple, why did *Adam* (rather than Eve, God, etc.) eat the apple, and so on. A functional hypothesis may explain perfectly well why a given trait persists rather than not (for example, after an organizational strategy arose, it allowed the organization to better attract resources than competing organizations). At the same time we may have no answer to the question why that strategy persisted rather than another (that could have arisen but did not). Two different questions are at issue here: Why, given that a practice arose, did it persist? and Why did one practice arise and not another? Because functional explanations may not answer the former, it does not follow they cannot answer the latter.

Some critics argue that functionalism trades on an illegitimate analogy with natural selection. Hallpike (1986) and R. Young (1988) argue that social entities do not reproduce; thus functionalism cannot be grounded on mechanisms analogous to natural selection. This criticism assumes not only that social entities cannot reproduce but also that they must for selection to operate. Neither assumption is reasonable. In some instances, social entities do in effect leave copies of themselves – Boyd and Richerson (1985), for example, describe a clear sense in which cultural practices "replicate." Furthermore, as Sober (1984) has pointed out, natural selection mechanisms do not require literal copying – only differential representation in future generations. Hannan and Freeman described a process of differential longevity and foundings for organizations, even though organizations do not literally make copies of themselves. Such differential representation suffices for selection to operate. Moreover, selection can occur even if birth rates are not determined by selective causes, so long as mortality is. Similarly, Ellen's (1982) worries that new practices do not originate in a random fashion are not decisive. Natural selection could still operate on variations that were somehow called forth by the environment – the latter process would then need explaining. Hannan and Freeman's differential survival is in the same boat, and they do have something to say about origins. So while Hannan and Freeman's work does not provide an exact parallel to natural selection processes, it does show that differential sorting processes can undergird functional explanations in the social sciences – contra the doubts of critics.[27]

Finally, doubts about cultural evolution need not be doubts about functional explanation. Many practices might be functionally explained without social change showing any overall direction. Evolutionism is added baggage to the functional program. Just as Darwinian functional analyses do not entail that biological change has any innate direction, so too for functional accounts in the social sciences.

Functionalist social science has long been a popular target for critics of the social sciences. If any kind of social research and theory looks beyond the scientific pale, it is seemingly functionalism. Yet even where the prospects for science in the social sciences seem most remote, good work is possible: functional explanations can be perfectly legitimate causal explanations, they are amenable to ordinary causal evidence, and they rest on no untoward analogies to biological evolution. Moreover, good work is not only possible but actual. The data sets, theoretical reasoning, and statistical methods of Hannan and Freeman compare quite favorably to what evolutionary biologists can muster. Thus skepticism about naturalism fails precisely at the point where it should be most compelling.

[27] For example, van Parijs (1981, p. 59), Elster (1983), and Rosenberg (1988, p. 136).

5

The failures of individualism

The last two chapters defended the prospects of a real social science. Yet the idea that social science can be good science is only half the classical tradition I want to defend here. That tradition argued not just for a social science, but a science of *society*. A social science was to be a science of the large scale – of patterns and processes explained in large part at the macrosociological level. I investigate and defend this tradition in what follows, arguing that good holist social science is both possible and essential.

We thus shift topics in this chapter. Nonetheless, the results of previous chapters will play a key role. The work of Paige and that of Hannan and Freeman, I shall argue, are instances of successful and irreducible holist research. Furthermore, defending holism will also be further defending naturalism. Individualists claim that the only adequate social science is one that eschews reference to macrosociological entities. Yet most social science ignores this demand. So if the individualists are right, there are serious doubts about naturalism. Rejecting individualism will at the same time remove one further obstacle to science in the social sciences.

The positive picture of the social sciences I shall defend is one that I believe is generally true of whole-part or macro-micro relations in the sciences. That view holds that wholes are, of course, composed of or exhausted by their parts and do not act independently of them; that, nonetheless, theories at the level of the whole can be confirmed and can explain at that level, without a full accounting of underlying details; that theories at the level of the whole may have only a messy relationship to how microlevel theories divide up the world, thus making macrolevel theories irreducible; that searching for lower-level accounts can be informative as a complement to, but not as a substitute for, more macro investigations; and that reduction is not the only route to the ideal of a unified science. This outlook is common in the philosophy of mind and the philosophy of biology.[1] My goal is to show that this view is equally compelling in the social sciences.

The individualism–holism debate has endured for centuries. Hobbes and Rousseau were early antagonists, and Durkheim and Weber continued the debate into this century. Current social science is equally divided on the issue: the rational choice approach, a major movement in the social sciences, raises the individualist banner while some sociologists deny that social science should study individual behavior at all (Mayhew 1980). Mainstream economics seems predicated on the idea of explaining large-scale economic behavior in terms of individual preferences and action. In fact, most systematic works in the social science begin by taking some stand on the holism–individualism issue.[2]

What makes this debate so enduring? Both sides can appeal to deeply held intuitions; both see moral implications riding on the debate's outcome. Individualists point out that society does not exist independently of its members and that social institutions do not act on their own. We may *say* that "the state usurped power" or that "too many dollars are chasing too few goods." Yet only the confused take this talk literally – it was particular individuals who took power and dollars do not have feet. So individualism seems like common sense metaphysics, holism like mystical musings.

Individualists also have strong moral intuitions on their side. If there are causal laws determining how institutions behave, then individual freedom, dignity, and responsibility are threatened. After all, the most famous holists have been dictators or their ideological inspirations. Marx and Communism or Hegel and fascism come quickly to mind when we mention such holist notions as collective interests or the laws of social evolution. Well-known individualists in this century – for example, Popper (1982) – explicitly defended individualism as an antidote to these immoral holist tendencies.

However, the holists can also appeal to some deeply held beliefs, especially from the Enlightenment and scientific traditions. Consider, for example, these ideas: (1) popular opinion is frequently riddled with superstition and misconceptions, and scientific progress requires going beyond common notions and engaging in rational investigation; (2) scientific explanation cannot stick simply with surface phenomena but must go behind the appearances; (3) science has shown that our world is a material

---

[1] See Fodor (1974) for a classic statement. For the case of biology, see Kincaid (1990d, in press a).
[2] See, for example, Coleman (1990).

world driven by material forces; and (4) good science deals with the repeatable – it produces repeatable results and finds repeating patterns in nature. Holists believe these facts about science support their view.

For example, holists deny that individualism's common sense appeal is reason to believe it. Rather, it is reason to be suspicious. Common sense physics is stuck in the Middle Ages; ordinary judgments on probability and numerous other scientific matters are notoriously bad. So why suppose people are any more experts on social matters, particularly when passionately held moral and political views lurk so closely in the wings? Yet individualist explanations appeal to common sense psychological categories like belief and desire. Furthermore, individualism takes individual beliefs and behavior at face value, since individualist accounts begin with these factors. Yet scientific progress has come from going beyond first appearances. Might not individual behavior have causes, constraints, and effects far beyond those apparent to common sense? Moreover, explaining behavior by appeal to ideas – beliefs and preferences – seems a holdover from prescientific days; materialism would look for concrete, quantifiable forces. Finally, individual behavior has numerous and diverse causes. If social science is to find repeatable patterns, it will have to be at the aggregate level where the idiosyncratic balances out. Thus a real science of society must study social phenomena in their own right – not rehash common, totally unscientific conceptions.[3]

Holists also have a moral ax to grind. Individualists may think they are defending human autonomy. Yet their freedom seems a particularly limited one to the holist. By explaining social phenomena from individual traits, individualists naturally ignore the real social constraints that limit human action. Poverty becomes the result of character flaws and unemployment the voluntary choice of workers. The real demands of justice – equal opportunity and access to social goods – go by the way. Individualism seems easily to become a morally insidious rationale for an unjust status quo. So the holist reasons. Thus the holism–individualism debate is not merely an academic debate over ontological commitment or the relation between theories.

Of course, the rationales cited above are fuzzy and of varying worth. This chapter seeks to turn these intuitive rationales into clear theses and arguments. Later sections will work hard to separate the many different theses involved. In broad terms, however, we can classify individualist claims as theses about either reduction, explanation, confirmation, or ontology. Reductionist theses claim that good macrosociological work can

---

[3] Arguments like these are advanced by traditional holists like Durkheim (1933) and contemporary holists like Mayhew (1980).

The failures of individualism

be eliminated in favor of individualist theories. Section 5.1 examines this version of individualism, arguing that the issue is ultimately an empirical one, with the evidence favoring the holist. Individualists also make many claims about explanation. Sometimes they claim that individualist theories can fully explain all social phenomena, even if reductions in the technical sense are unavailable; or, more strongly, they claim that individualist explanations are always the best explanation. Others assert that individualist mechanisms are necessary for explanation or confirmation. Section 5.2 takes up these and related theses. Section 5.3 applies the results of these discussions to individualism as an ontological thesis. Finally, Section 5.4 concludes by summarizing the real insights behind the individualist position.

5.1 The prospects for reduction

Individualists frequently claim that all good social theory is reducible to individualist accounts. Though there are many other individualist doctrines, we will see in later sections that these other varieties often have close ties to their reductionist cousin. Reductionist versions of individualism also have clarity on their side. Philosophers have studied the requirements for reduction in detail as well as the potential obstacles that confront reductionist programs. Hence I start this chapter with the claim that social theories are reducible. After describing requirements for reduction, I consider and reject conceptual arguments claiming to decide the issue. Reducibility is an empirical matter, and I try to make an empirical case that (1) alleged instances of reduction fail and (2) that there are well-confirmed social theories that are irreducible.

5.1.1 *Requirements for reduction*

Individualists and holists use the term "reduction" loosely. Yet we need a clear account of reduction to evaluate this version of individualism. This section therefore outlines the requirements for reduction.

The root idea beneath reductionism is that one theory is a special case of another. Showing that one theory is a special case means showing that we can derive its results from some more fundamental theory. This practice is common in science – Newton showed Kepler's laws to be a special case, and Einstein did the same for Newton's laws. Particularly powerful reductions come when the macroscopic is derived from the small scale, as perhaps has happened with thermodynamics and statistical mechanics. Individualists claim something similar can be done for social theories.

How do we separate the *social* or *holist* theories from the *individualist* theories that are to reduce them? Social theories refer essentially to social entities or events and their properties – for example, to social institutions and classes or to social events such as revolutions or depressions. However, individualist theories might allow quantification over sums of individuals and thus they might also refer to social entities, if we can identify the latter with such sums. So we might alternatively pick out holist theories by their language. Social theories are ones that use social predicates or kind terms such as "class," "organization," "state," and so on.

Individualist theories must refer only to individuals. Yet "referring to individuals" hides important ambiguities. The description "is a member of the ruling class" refers to an individual, but it will not do for individualism. It invokes a social entity, the ruling class, and thus would not eliminate social explanations – the goal of reduction. So an individualist theory must be purely individualist; it must refer only to individuals and make no ineliminable assumptions about social entities or ineliminable use of social predicates. This does not commit the individualist, however, to the radical view that we can explain social events without reference to relations between individuals. While some individualists hold this view, it would be unfair to saddle all with such an implausible doctrine. So long as the relationships involved make no assumptions about social entities or avoid social predicates, no harm is done. Of course, the distinction between social and individual relations or predicates may sometimes be unclear. I have no automatic procedure for distinguishing them; they must be approached case by case. The best argument against individualism will let questionable cases in; the weakest case for individualism will do the same.

It is fairly clear what reduction does and does not require. First, what it does not: Reducibility does not demand that actual reductions be on the books, only that they be possible in principle. Of course, the best way to establish reducibility in principle is in practice. Individualists can appeal to what might be done in the future only so long; at some point they must pay up. Reducibility also does not require that weak, poorly established hypotheses be reduced; it is only *good* macrosociological results that must be eliminated.

Now for what reduction does require. To reduce one theory to another, we must first have some way to tie the two together, for they describe the world in very different vocabularies. So the first condition on reduction is that we have some way to translate a holistic theory into an individualistic one by substituting the basic descriptive terms of the individualist theory for those of the holist theory. This is sometimes called the connectability requirement. However, not just any translation of social theory into individualist theory will do. Our goal is to capture the explanations of social theory in individualist terms. We can only do that if our translation allows us to derive the laws or other explanatory statements of the holists's ac-

The failures of individualism 147

count. In short, to reduce we need (1) some way to connect the descriptions of social theory to those of individualist theory and (2) a connection that, together with other information, allows us to capture the explanations of the social theory.

What kind of connection do we need then between the basic terms of social and individualist theory? Let me say first what we do not need: a complete equivalence of meaning.[4] The notion of synonymy hardly wears its meaning on its sleeve, as we saw in Chapter 2. We also saw in Chapter 2 that we can compare theories without perfect equivalence of meaning. Thus the connection I shall adopt here is the notion of law-like coextensionality. Law-like coextensionality requires that social terms and individualist terms have the same application or reference; each social term always holds of the same entities described in individualist terms. More precisely, we need a law-like biconditional connection from every social term to some individualist term.[5] These connections are typically called "bridge laws," since they bridge the gap between lower- and higher-level terminology and allow us to take each social description and substitute some individualist description.[6]

While law-like biconditionals between lower- and higher-level terms are necessary for reduction, they are not enough for two reasons:

[4] Lessnoff (1974) seems to accept synonymy as a requirement for reduction (and thinks it can be met).
[5] Contra Sensat (1988), it is quite obvious that we do not need a connection running the other way. Some individualist descriptions may be of behavior that has no social counterpart.
[6] Richardson and others have argued that biconditional bridge laws are unnecessary – that all is needed is that "each physiological type [individualist type in this case] . . . should map onto a psychological type [social in this case]" (1979, p. 548). Richardson's claim is confused for several reasons. A conditional from each lower-level type is compatible with some higher-level types being left out entirely; Richardson has confused the scope of his quantifiers and should say that for each higher-level predicate, there is some lower-level predicate from which there is a mapping. More importantly, this weaker requirement does not suffice for reduction. A simple conditional from lower to higher does not ensure that we capture the laws of the higher-level theory when there are multiple realizations, for the higher-level predicate in a law refers to all realizations, something no single lower-level predicate does unless that predicate conjoins all possible realizations. But if we can list all possible realizations, then we *do* have a biconditional bridge law. One-way conditionals also do not suffice for reduction because reduction requires more than simply connecting up terminology – it requires doing so in a way that allows us to capture higher-level explanations soley in lower-level terms. As I point out below, *biconditional* bridge laws do not guarantee this, so clearly the weaker requirement will not suffice either. There is a sense in which Richardson's weaker conditional does allow case-by-case lower-level explanations, but those explanations are quite anemic, as I point out in the next section.

(1) Biconditional bridge laws are compatible with a many–many relation between higher- and lower-level terms. The many–many relation I have in mind might be pictured like this:

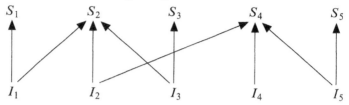

The S's are social descriptions or categories and the I's are individualist categories *and* this situation holds at a particular instance. Note that we still have biconditional bridge laws – for example, whenever I_3 holds, so does S_2 and vice versa. Something like this relation probably holds between genes and DNA or beliefs and neurological processes. In both cases, we have two complex systems at different levels that do not match up – even though one is realized by the other. For example, my belief that London is a wonderful city probably involves various neurological connections that do more than just process thoughts about London. If this many–many relation holds, then we have no way of equating kinds or descriptions one by one. That means we cannot reasonably substitute individualist descriptions for particular social ones; the explanation for any particular social process is always the same: the totality of individual processes. Reduction fails.

(2) Biconditional bridge laws by themselves are also not enough for a second reason. We want our reductions to eliminate social theory – to show that individualist theory completely captures true social explanations and thus that the latter are no longer needed in principle. Yet a lawful connection between the categories of the two theories does not ensure that social theory has been eliminated. Our individualist accounts of the world might *presuppose* social theory by using social variables in one way or another. If they do so, social theory is not completely reduced.

Thus bridge laws by themselves do not suffice for reduction, despite the common misperception that they do.[7]

---

[7] Found, for example, in Kim's (1989) interesting attacks on non-reductive physicalism. It is surprising that this point has gone largely unnoticed, since it seems an instance of the well-known problems with nomological-deductive accounts of explanation.

The failures of individualism

So what exactly does reduction require? I will skirt a complete account here, for our argument against individualism can get by with less. I have argued that reduction requires (1) law-like one-to-one mappings from social predicates to individual ones which (2) allow us to capture each true social explanation without (3) presupposing social theory. It is these requirements we shall use in what follows to evaluate reductionist forms of individualism. It is important to note, however, that these conditions describe reduction in its pure form. Actual scientific reductions are often a considerably more messy process. Some parts of the theory to be reduced are corrected by the reducing theory as partly inaccurate; other parts may be dropped altogether because they simply represent confusions. So in actual practice there is a continuum with reduction at one end and elimination at the other. Sometimes theories to be reduced are largely shown to be false, with only small parts being actually reduced. In this section I focus on the pure case of reduction. However, later sections on explanation and confirmation will answer theses that argue for reduction via elimination.

5.1.2 *Conceptual arguments for and against reducibility*

Philosophers and social scientists have typically tried to settle the individualism–holism dispute on a relatively *a priori* basis. Both sides try to provide broad conceptual facts that would show reduction possible or impossible. This approach continues unabated up till the present (see, for example, James 1984). Since these arguments dominate the field, I consider them next. However, previous chapters should make it abundantly clear that I find this approach wrong headed.

Holists typically believe there are true causal laws connecting social entities. Unless they believe that social entities exist independently of individuals, they then also believe that social entities cause individual people – in brief, holists are committed to "downward causation."[8] To individualists, however, deterministic social causation is both conceptually confused and incompatible with a deep truth about human beings. The conceptual confusion comes from having wholes influencing their parts. Since wholes consist of and do not act independently of their parts, they cannot also cause them. Furthermore, causally efficacious social entities are incompatible with human free will. If laws govern the social world, free will goes out the window. Yet free will seems a far more certain truth than does holism; so it is the holism that should be tossed.

Neither argument goes anywhere. Let us look first at the claim that downward causation is conceptually incoherent. Individualists worry that it conflicts with the commonplace that wholes do not act independently

---

[8] For example, Agassi (1973) makes this charge.

of their parts. Does this principle really conflict with downward causation? It is easy to see it does not. If some state of a whole W, W_s, causes one of its parts, then whatever realizes or brings about W_s also causes that part. So downward causation simply implies that some configuration of individuals at one time causally influences one of those individuals at a later time. Surely no incoherence lurks here.[9]

Social causation certainly is incompatible with freedom defined as "uncaused." Yet this libertarian notion is notoriously problematic. If individualists want freedom in this sense, then causation at the *individual* level is ruled out as well. So even an individualistic social science goes by the way. On the other hand, suppose we take freedom as do compatibilists, as acting on desires and preferences free from coercion. Causally efficacious social wholes are then thoroughly compatible with freedom so defined. Wholes stand in causal relations to each other and to individuals. Yet that causal influence comes about through the actions of individuals, who may be acting on their desires and preferences. In short, if freedom is compatible with causation, then social causation is compatible with freedom.

Holists sometimes try to use downward causation for their own ends. If social wholes cause individuals, then holists argue that social theory is ineliminable. No fixed, universal nature of individuals will suffice to explain all social facts, because individual behavior itself depends on social context. This reasoning occurs again and again in the literature.[10] The individualist is committed, the reasoning goes, to an explanatory chain where every social process, S, is ultimately preceded by only individualist explanations. For example, individualists can allow a chain like the following where the social explains the individual:

$$S_1 \longrightarrow S_2 \longrightarrow I_1$$

only if the full chain becomes:

$$I_3 \longrightarrow I_4 \longrightarrow S_1 \longrightarrow S_2 \longrightarrow I_1$$

However, social facts apparently are always involved in explaining individual behavior; the bare traits of human nature do not suffice.

Individualists have a ready response to this argument: the influence of social context can itself be explained in terms of individuals. After all, temperature influences the molecules in a gas. Yet it would be silly to think this shows that the gas laws are irreducible to statistical mechanics, for

---

[9] This argument of course turns on assuming that social entities or states are token identical with some complex state of individuals. Some holists or metaphysicians will no doubt try to argue against this claim. Yet individualists obviously will not, and they are the target of the argument.

[10] See Ruben (1985, ch. 4).

temperature has a statistical mechanical explanation. Thus the macroscopic gas laws are not irreducible simply because macroscopic features influence molecular ones. The holist has at most identified a potential problem. To make a real case, the holist has to show that actual individualist explanations cannot reduce the presupposed information. Simply pointing out that social factors influence individual behavior will not do.

We should note in this connection that individualists need not endorse a fixed human nature or trans-social principles of human behavior, as holists often charge.[11] Basic human traits and dispositions can vary according to context, even social context, so long as we can cash out the context in individualist terms. Even holists sometimes hold that social laws are contextual; Marx, for example, defended what has been called "historical specificity" (Korsch 1938) – the thesis that different social formations have different laws of behavior. If holists can avail themselves of such specificity, then so can individualists.

Individualists have also thought any social theory reducible because of several undeniable truths about social events. Watkins, for instance, cites two "metaphysical commonplaces" supporting reducibility: (1) "the ultimate constituents of the social world are individuals" and (2) "social events are brought about by people" or "it is people who determine history" (1973, p. 179). These commonplaces, Watkins claims, have the "methodological implication that large-scale phenomena ... should be explained in terms of the situations, dispositions, and beliefs of individuals" (1973, p. 179). This methodological injunction, of course, presupposes that reduction is in principle possible (or that large-scale phenomena cannot be explained at all, a view Watkins does not hold). Thus, reducibility allegedly follows from the fact that societies neither exist nor act independently of individuals. This reasoning occurs again and again in the individualist literature.[12]

We can spell out Watkins's two commonplaces in terms of exhaustion and supervenience, ideas developed to defend physicalism or materialism in the philosophy of mind. Physicalism, as an ontological view, makes two basic claims: (1) that physical entities *exhaust* what there is – everything that exists is a physical entity or a complex sum of physical entities – and (2) that the non-physical facts *supervene* on the physical. "Supervene" is a philosopher's word that can be quite precisely defined (Hellman and Thompson 1975), but for our purposes this rough formulation will do: a set of facts A supervenes on the set B if once all the B facts are fixed, then so too are the A facts. In philosophy of mind, for example, this thesis

---

[11] As claimed by Sensat (1988, p. 195).
[12] For example, by Collins (1981, p. 989) and Mathien (1988, p. 11).

would hold that two individuals physically identical in every respect would also have all the same beliefs, desires, and other mental states. Applied to the social realm, these two physicalist ideas give some more content to Watkins's commonplaces.

So Watkins's claim is that exhaustion and supervenience entail reducibility. Watkins is wrong. Neither alone nor jointly do these principles ensure reducibility. A social entity S may be composed entirely of individuals $I_1 \ldots I_n$ and not act independently of them, yet that tells us nothing about the power of our theory of $I_1 \ldots I_n$ to capture higher-level explanations. Supervenience does not automatically bring reduction for reasons we have already implicitly given. Supervenience only requires a one-way mapping from each individualist predicate or term to some social term or predicate. Yet a mapping from each individualist description to each social description does not ensure the one-to-one mapping that reduction requires. For example, the same social process might supervene on or be realized by indefinitely many different individual-level processes. This possibility of "multiple realizations," as it is called in the literature of philosophy of mind, means we may not have the bridge laws needed for reduction – even if social processes are exhausted by and supervene on the actions of individuals. Moreover, we saw in Section 5.1.1 that (1) individualist accounts might presuppose social accounts and (2) the relation between individual and social facts might still be many–many, even if we have one-to-one mappings from the individual to the social. Both situations thwart reduction. Yet nothing in Watkins's commonplaces *entails* that these circumstances cannot obtain, again showing that reducibility is not guaranteed.[13]

I want to consider one last conceptual argument concerning reduction, this time from the holist side. Ruben (1985) argues that a basic truth about rational or justified belief shows individualism doomed. A belief is rationally held, says Ruben, only if it is in part explained by what it is about. For example, your belief that you are now reading a book is rational only if there is a book in front of you causing that belief. However, among our beliefs are some about social entities. So if my beliefs about social entities are justified, then social entities must partially explain my belief. Thus individualism cannot be right. In Ruben's words, "If our beliefs about society . . . cannot be explained in terms of the social facts of the matter, then whatever does explain them will not confer rationality on them" (1985, p. 170).

---

[13] Some philosophers (see Rosenberg 1985) still try to squeeze reduction out of supervenience by claiming that we can always form reductive definitions by disjoining the relevant supervenience bases. This won't work, for there may be indefinitely many supervenience bases unknown to us and because even if we know all the supervenience bases our descriptions of them may still be in higher-level terms or their relation to the reducing theory may still be a many–many relation. See Kincaid (1987, in press a).

The failures of individualism

Ruben's argument does not succeed. We should note first that his account of justification is not uncontroversial. Coherentists will deny that there actually must be a causal process from the external world for my belief to be justified. If I am being systematically deceived in ways I could not detect, then I can still be rational so long as I use my best evidence, rules of logic, and so on. More important, even if we accept the causal requirement for justification, Ruben's conclusion cannot follow. Imagine a parallel argument given by the friends of the gas laws: I have rational beliefs about temperature, pressure, and volume. Therefore we cannot explain these phenomena in molecular terms. Surely something has gone wrong somewhere. The problem seems to be this: Individualists can allow that social phenomena explain our beliefs. They can do so *as long as* they can show that those social explanations are really special cases of individualist theories. And that is just what they claim. Ruben has at best pointed to a potential obstacle to reduction.[14]

So we have examined enough conceptual arguments. Though there are many more, we can safely ignore them. We have good Quinean grounds (and good inductive ones as well!) to do so. No broad conceptual argument is going to tell us what science can or cannot ever do. Individualism must be defended – and criticized – on empirical grounds. That is the task of the next section.

5.1.3 *An empirical case against individualism*

A strong case for individualism will show that good social theory can be reduced to good individualist theory. Since conceptual arguments will not achieve this result, individualists have to look at actual cases. They need to take real social theories and convincingly argue that reduction is possible. They can best do that by providing at least partial reductions – reductions of some important parts of our best social explanations. Holists, of course, need to do the opposite, namely, take actual holist theories and show that reducibility is unlikely. Holists can establish this negative claim in two ways: (1) by showing that, based on what we know now, the potential obstacles to reduction are real and (2) by arguing that alleged individualist reductions fail.

Potential obstacles to reduction. Numerous potential obstacles confront reduction.[15] Recall that for reduction we need (a) biconditional bridge laws connecting the basic predicates of social theory to our best individualist theory (of the form S iff I, where S is a social term and I an individualist

---

[14] To be fair, Ruben may intend to establish no more than this.
[15] For an early and particularly helpful description of these issues, see McCarthy (1975).

one) such that (b) we can derive all social explanations from individualist theory without (c) presupposing the theory to be reduced. These requirements can fail because of the following problems:

> Multiple realizations: If a social description applies to phenomena that can be brought about in diverse ways by individual behavior in different situations or at different times, then serious difficulties arise. Basically our coextensional bridge law fails in the S-to-I direction. As we will see later, multiple realizations are particularly likely in the case of high-level causal claims that abstract from individual detail or when we define terms by their role vis-à-vis other social factors.
>
> Presupposing social information: Even if we can supply a coextensive connection between higher- and lower-level terms, we are not yet out of the woods. Individualist theories might still explicitly or implicitly presuppose truths about social entities, events, and so on. We briefly saw above how this might happen: individuals are described as, say, members of a particular class or as holding a position in a particular institution. Such descriptions arguably use social theory rather than eliminate. Alternatively, explanations of individual behavior may quite directly invoke macrosociological variables. In either case the resulting explanation does not replace social theory with an individualist account, the goal of reduction.
>
> Many-many relations: Social kinds or terms and individualist kinds can stand in a messy many–many relationship, even if we have biconditional bridge laws. That means we cannot capture specific explanations and reducibility in any meaningful sense fails.
>
> Context sensitivity: Context sensitivity occurs when a particular individualist description maps onto different social descriptions depending on social context. If that happens, then our biconditional bridge law fails in the I-to-S direction. Context sensitivity is most likely when the social significance of behavior is what counts. For example, an individual act of violence may not uniquely determine its social description until further social context is provided. A worker shooting the foreman may be an act of religious hatred in Northern Ireland and racial conflict in South Africa. If the individualist description only determines its social description with the help of other facts, then either the individualist description fails to provide the appropriate one-to-one mapping or it presupposes social theory, thereby not sufficing for reduction in either case.

The failures of individualism

These potential problems all come to one simple point: the social sciences and the sciences of individual behavior may divide up the social realm in very different ways, ways that bear no neat relationship to one another. This is a common sense moral we drew in Chapter 2 from Kuhn's claim that different theories are incommensurable. While Kuhn thought such incommensurability inevitable, that is not my argument here. The obstacles sketched above are only possibilities, and a convincing case against reductionism must show they are empirically real. I turn to that task in the next two sections.

Failed attempts at reduction: rational choice theory. Rational choice theory is probably the most developed and sophisticated approach allegedly supporting the individualist program. Rational choice theory starts from individuals and their preferences and then builds up explanations of social phenomena. The basic approach, which goes back as far as Hobbes, works this way: (1) assume individuals are rational, self-interested, and acting under constraints, and (2) explain social phenomena by deducing them from these assumptions about individual behavior. In this section I look at rational choice theory and ask whether it supports individualism in its reductionist guise. Since the rational choice approach covers an enormous amount of work, I will focus on three of its seminal defenders: Homans, Becker, and Downs. All three, I shall argue, fail to support the individualist position.

George Homans (1974) was an early advocate of individualism. He consciously tried to develop theories illustrating the individualist program. More influenced by behaviorism than decision theory, Homans sought to show how the dynamics of small groups could be reduced to individual dynamics. Among Homans's main results are claims like the following:

(1) Groups control their members by creating rewards they can withdraw.
(2) Groups ostracize deviants.

Simply put, group dynamics result from the incentives and disincentives facing individuals, who seek to maximize their "psychic profit."

Though in one sense Homans has given an individualist explanation, he has not shown that group laws are reducible. His alleged reduction – "social groups control their members by withdrawing rewards" – faces three of the obstacles to reduction discussed earlier:

> Multiple realizations: There will be innumerable social arrangements for rewarding and punishing. Material incentives, social esteem, religious salvation, political power, and so on all can do the job, and do it in various ways. No single, fixed descrip-

tion of what individuals do to bring about "group rewarding" seems available. For example, we cannot group these various realizations simply by whether individuals are rewarded by their interactions, for not all rewarding interactions between individuals – asocial behavior, for example – realize group rewarding.

Context sensitivity: Not just any rewarding activity brings about "group rewarding"; whether a relation instantiates this predicate will depend on the social context. Some rewarding personal interaction will in some circumstances have nothing to do with group control or will even be counterproductive.

Presupposing social information: Explaining which personal interactions do count as "group rewarding" will naturally lead us to invoke the social roles which individuals play. Rewarding interactions between individuals will contribute to *group* control only if they bear some relation to group structure. But one natural way for that connection to be made is by way of social roles – interactions that reward as teacher, inmate, or employee. However, these designations seemingly presuppose social truths – about schools, prisons, and the like. In fact, other sociologists (Bates and Harvey 1975) have raised criticisms of individualist small-group theory, based in large part on this point and the preceding one.

Thus, while individual interactions, perhaps even ones based on psychic profit, may bring about all social phenomena as Homans claims, reduction seems unlikely. Note, furthermore, that the social phenomena Homans discusses are relatively small scale – precisely where reduction should be most likely if possible at all.

Downs and Becker develop their theories with far more formal rigor than Homans. They also drop the behaviorist underpinnings, opting instead for the standard economic model of utility maximizing under constraints. Both took the novel step of applying standard neo-classical tools to non-economic social behavior. We cannot deny the creativity, fruitfulness, and insights that may have come from this extension. Yet I do intend to deny that these theories support the reductionist form of individualism.

Becker (1976a, 1981) uses the economic approach to explain discrimination, crime, the family, and social interaction more generally. In each case the approach is the same: assume individuals maximize their fixed preferences as best they can. In the case of discrimination, Becker assumes that whites have a "taste for discrimination" and then asks how that will affect ordinary market behavior. Discrimination means that the two groups trade less with each other than would be optimal under free trading. Both black labor and white capital suffer. Blacks face a more restricted capital market, white capital a more restricted labor supply. Given that blacks are in the minority, they will be hurt more than whites. By making some further assumptions about production functions, Becker estimates that black income should be between 39 and 55 percent of that of whites. The observed figures do fall into this range. Thus Becker (1976a, p. 24) concludes that

> market discrimination against Negroes in the United States could easily result from the manner in which individual tastes for discrimination allocate resources.... this implies that monopolies, political discrimination, and the like, are, at most, secondary determinants... and that individual tastes for discrimination... constitute the primary determinant.

Becker analyzes the family in a similar way. The family is a production unit. It combines individuals and their time, traits, resources, and preferences, and produces an aggregate commodity which we might call, after Tolstoy, "family happiness." It is the aggregate of income, love, children, and caring that households produce. Individuals enter into households because and when their total goods from living alone are less than those from marriage. Becker notes that individuals living alone cannot substitute their time in the form of wages for consumption of other family goods; there is no efficient market in which individual households can purchase caring, children, or love. So the goods individuals have to offer are complements; individuals will be better off forming a union than they would be trying to get these goods in the market.

This account entails several things about families. Monogamy is generally more efficient, since adding more spouses has diminishing returns. Individuals will differ in the number and quality of children they wish to produce; their preferences will influence the timing of marriage and potential marriage partners. So, too, will the preferences for beauty, intelligence, and other traits of potential mates. Likes will tend to marry likes, because most marital goods are complements – and each individual will gain more by contracting in marriages to those with complementary traits than with those whose traits are substitutes. Becker provides more detail, but these results give the flavor of his work.

We may wonder whether Becker's model picks out real causes, since it rests on numerous untested *ceteris paribus* assumptions and since he does not rule out competing explanations. However, my most direct concern is whether Becker's models support individualist reductionism. I think they do not for several reasons. First, Becker's theory does not eliminate social explanations because it assumes them at every turn. Consider the following factors which Becker calls exogenous – that is, explained by outside factors:

> individual preferences for number of children and their "quality"
> individual preferences for traits of spouses
> relative wage and property shares of males versus females
> the taste for discrimination
> the relative shares of capital owned by the races

There is nothing intrinsically wrong with leaving these variables as exogenous. Yet doing so is nearly fatal if we want to use rational choice theory to reduce.[16] All these "exogenous" factors are precisely ones that we would expect to explain *holistically*.[17] Attitudes about children, about sex roles, and about racial prejudice seemingly have their roots in larger social structures – the family; the media; peer, work, and ethnic groups; religious affiliation; and governmental and judicial institutions. Relative property and income shares also depend on social factors – historical factors like slavery and patriarchy. In short, the exogenous factors all seem to be social. Rational choice theory seems to presuppose rather than eliminate social theory.

Of course, the rational choice advocates *could* try to explain these exogenous factors in individualist terms. Yet that is at best a hopeful promise. The rational choice approach is really neo-classical economics transported to non-economic phenomena. However, neo-classical theory has no account of preference formation; it, like Becker, takes preferences as given. So there is no well-developed tool for individualists to pull from the neo-classical theoretical shed. Much the same holds for assumptions about relative shares. Neo-classical theory does have accounts of income determination; it also has growth models, though those are much more contentious. Yet neither of these components gives the whole story about who gets what. Standard marginal productivity theory *starts from* a given distribution of wealth and preferences and then explains the resulting market allocation. However, assuming a prior distribution of wealth and preferences is again leaving room for slavery, military power, robbery and corruption, political interventions, and other social explanations of wealth

---

[16] I do not recall Becker specifically espousing individualism. Yet others – for example, Elster (1985) and Hechter (1983) – certainly do cite it as supporting the individualist program, and Becker is that approach's seminal and best-known advocate.

[17] This point is made by Hindness (1988) and Mayhew (1980).

The failures of individualism

differentials. And if we look at Becker-style accounts of individual earnings – human capital theory – the empirical evidence makes it clear that social variables must be included.[18] So rational choice theory is not a purely individualist theory.

There is another reason why Becker's rational choice theory does not support reduction: it is likely to run into the multiple realizations problem. Discrimination, for example, may result from quite different combinations of individual preferences and initial endowments. While rational choice theory might be able to explain each case, no reduction would be forthcoming, because the social process as a kind of event has no equivalent in individualist terms. So to the extent that we have explanatory social generalizations about discrimination, they will be irreducible. Similar conclusions should hold for all social phenomena that can be brought about by diverse distributions of preferences.

Downs's (1957) theory of political behavior likewise imports neoclassical results into a non-economic area. Since the basic outlines parallel the work of Becker – who came up with the same ideas independently – I shall briefly state Downs's results and the corresponding problems. Downs's theory assumes that voters and political parties are maximizers. Parties try to maximize votes, at least up to a point; individuals vote for those parties that will maximize their own well-being. Since gathering information is costly, voters will look for simple ways of identifying the right party. Parties will oblige by adopting "ideologies" – slogans that voters can easily grasp and identify with. For voting populations spread on a left-to-right continuum, interesting results about party position follow. The situation is like that of two stores trying to decide where to locate in an evenly dispersed neighborhood. Given some other assumptions, Downs shows – borrowing results elsewhere – that positions near the center are optimal. He produces other interesting and intuitive results, but this sketch suggests how they go.

Downs, like Becker, does not make a serious effort to confirm his *ceteris paribus* assumptions; furthermore, his selection of data is not entirely consistent with the best evidence (Barry 1978). More important for my purposes, Downs's work – and I would argue that of the whole public choice school in political science – does not support a strong individualist position, for the same reasons that Becker's work does not. The political preferences of individuals are exogenous. To solve collective action problems, we assume that individuals are pressured by or feel some solidarity with

---

[18] Variables like socioeconomic status, union membership, etc., are typically used when human capital accounts are empirically specificed. For a review, see Blaug (1976).

larger groups. Political parties and judicial structure are assumed. Thus even if confirmed, Downs's analyses work only against a background of social entities and regularities.

I obviously have not surveyed all rational choice theory. But the problems of Becker and Downs are not unique to them. Rational choice theory begins from preferences and the constraints individuals face. Both factors typically bring in social institutions, either implicitly or explicitly. Similarly, large-scale social processes may be the end result of quite different individual motives and beliefs. So the multiple realizations problem is probable. These facts are likely to confront any version of rational choice theory.[19]

Though the work of Downs and Becker is seminal, it is not the latest word in the rational choice approach. Many exciting results have recently come from game theory, a close cousin of the work discussed above but one that is often more precise and that pays more attention to dynamics. An extensive discussion is beyond my scope here, but I want to indicate briefly why game theory confronts precisely the same problems as do Downs and Becker. Game theory explanations are parasitic upon prior social information at every turn. For example:[20]

(1) Game theory analysis depends on precise rules of the game; without them it is very hard to derive any results at all about what outcomes are likely. Feasible strategies and strategy sequences have to be set before useful analysis can begin. Yet these rules of the game are just what the institutional background provides, and the institutional background, of course, is likely to involve social processes. Paige's peasants could be analyzed by game-theoretic notions, but the crucial institutional structure is one rich in social factors such as classes, states, and so on.

(2) Game-theoretic analyses frequently allow "multiple equilibria" – multiple outcomes given the initial set of strategies and payoffs. Thus game theory cannot fully explain what happens in reality; some other factors are operating. Norms, conventions, access to information, and so on are likely candidates. It would not be surprising, however, that these factors involve non-individualist explanations.

(3) Players' payoffs are generally taken as given and they are usually assumed to be monetary in nature. This means that game theory,

---

[19] See, for example, Hechter (1983) and North (1981), both of which run into similar problems qua individualist theories.

[20] For a survey of the field that highlights these factors, see Kreps (1990).

like rational choice theory, takes preferences as given rather than explaining them. This again opens the door to social factors. When payoffs are not assumed to be monetary – when beliefs about fairness and so on are allowed – game theory has a much harder time producing definite results. This further suggests that social factors are presupposed.

(4) Distribution of initial resources likewise makes a difference to the outcome of games. Collective action, for example, is not necessarily defeated by free riding when resources are differentially distributed.[21] As I argued earlier, this means that a background of social factors and entities is being presupposed, not explained.

Thus rational choice theory and its permutations do not provide us with individualist reductions. These approaches have contributed much, but their contribution is not to the reductionist program.

Some irreducible social theories. Some prime candidates for individualist reduction thus fail. To that extent holism is supported. Yet we can make a much stronger case for holism by showing that there are irreducible, well-confirmed social theories. The well-confirmed social theories I have in mind are those discussed in earlier chapters. Both the work of Paige and that of Hannan and Freeman are unlikely to be reduced, because the potential obstacles to reduction are real.

Let us look first at Hannan and Freeman's functionalism. Hannan and Freeman's work provides numerous causal generalizations about organizations, populations of organizations, and the social environment. Any attempt to reduce those generalizations will run into numerous obstacles, most obviously the multiple realizations problem. Two factors make multiple realizations likely: abstracting from individual-level detail and defining key terms by their social roles. Hannan and Freeman, however, develop a theory that invokes social entities at multiple levels of abstraction; their account also describes general *roles* social entities play vis-à-vis each other. Thus we should expect that diverse individual behaviors can realize their generalizations. That is indeed the case.

Recall from Chapter 4 Hannan and Freeman's various generalizations about organizations and the descriptive terms they employed. Among the basic predicates in Hannan and Freeman's account are:

---

[21] See Marwell and Oliver (1993) for a discussion of this and other social factors that play an essential role in more realistic rational choice and game-theoretic approaches to collective action.

"is reliable"
"is legitimate"
"is re-organizing"
"is an institutional norm of kind X"
"is a specialist"
"is a generalist"
"follows an r-strategy"
"follows a k-strategy"
"is a coarse-grained environment"
"is a fine-grained environment"
"is a political constraint on resources"
"is an organizational constraint on resources"

These terms can be brought about in indefinitely many ways by individual behavior. One way to see why comes from noting that Hannan and Freeman's theory of organizational diversity transfers the concept of fitness to the social level. The predicates listed above describe components of organizational fitness – those factors that determine which kinds of organizations die or thrive. It is widely recognized, however, that biological fitness is brought about by indefinitely many physical traits (Sober 1984, ch. 1). So we should not be surprised that social equivalents of fitness also have indefinitely many realizations in lower-level terms.

If we look at these terms one by one, that is exactly what we see. Take, for example, "specialist" and "generalist." Innumerable different specific organizational strategies will fall under those rubrics – especially since these terms apply across very different kinds of organizations. Moreover, even knowing all the ways organizations can be specialists or generalists would not end the problem. "Organizational strategy" is still not an individualist description. Different organizations will no doubt pursue roughly the same strategy in different ways – with different internal structures, incentive systems, control structures, and so on. And it is not obvious that even these latter descriptions are immune from the same problem; for example, we saw earlier that "the group rewards" is also prone to open-ended individualist realizations.

The obstacles do not end here. Suppose we had a complete individualist description that captured in a law-like way "is a generalist organization." Our job would only be half done, for we would still need an individualist account of the social environment. Yet Hannan and Freeman's environment is a *social* environment. It is constituted from political entities (states, judicial systems, government bureaucracies, and so on), social norms, economic entities, and other populations of organizations. So the individualist would also be forced to reduce these basic kinds as well. Yet

The failures of individualism

these social terms also describe higher-level processes that abstract from individual particulars. States, judicial systems, and non-profit organizations come about by many different configurations of individual behaviors and relations. The multiple realizations problem seems ubiquitous.

Given this fact, the other obstacles to reduction are also likely. To explain organizational behavior in individual terms, we would naturally invoke the basic roles individuals play. Yet invoking roles in institutions easily leads us to presuppose social information rather than eliminate it. Given the many levels of complexity involved in Hannan and Freeman's account, we can also imagine that similar individual behaviors will not uniquely determine their organizational implications. So context sensitivity may be likely as well. Since the multiple realizations problem is so overwhelming, I will not pursue these potential problems further.

So Hannan and Freeman's explanations are unlikely to be reduced to individualist terms. Since we have already argued that those generalizations are relatively well confirmed, we have a positive case of irreducibility. We can argue a similar claim about Paige's research, and I turn to make that case next.

Paige's theory, like that of Hannan and Freeman, is through and through a social theory – it explains via social entities and their characteristics. And once again the multiple realizations problem is endemic. Paige's basic terms or categories have intimate connections to one another; their role vis-à-vis other social entities plays an essential part in their definition. These social descriptions likewise abstract from individual detail. Not surprisingly, they apparently can be brought about in indefinitely many ways.

Among Paige's basic categories are:

>   urban elite
>   multinational corporation
>   bureaucratically administered organization
>   politically dominant class
>   financial institution
>   upper-class power
>   political power
>   judicial system
>   political repression
>   revolutionary movement
>   nationalization
>   nation-states
>   serf
>   union
>   urban political party

>political organization
>upper class
>class opposition to free markets
>political system
>revolution
>reform movement
>state

Many, perhaps all, of these terms have open-ended realizations in individual behavior. The *state,* for example, is a social entity, one which is surely defined by its social role and properties. It is, at least in part, the entity which has more control over organized violence than any other institution. Moreover, states are made from widely varying organizational structures. That means we might not even be able to reduce laws about states to ones about smaller organizations. If that is the case, then reducing talk about the state to some fixed and limited set of individual behaviors seems even more unlikely. Recall, also, that reductions must not cheat – they cannot refer to individuals in such a way that social explanations are presupposed. So individualists cannot group diverse individual behaviors by their social function. Moreover, if "state" is hard to replace with some individualist description, that holds equally well for the things states *do* such as "repress" social movements or favor particular classes.

Similar problems show up when we try to eliminate explanations citing the acts of upper classes. Such classes, in Paige's theory, are sometimes politically dominant; they also sometimes actively oppose free markets and other economic policies not in their perceived interest. Individualists are unlikely to know all the ways individuals can bring about these processes, for several reasons. For one, being politically dominant requires having disproportionate influence on the state. Yet, as we saw above, the prospects for eliminating that latter description are nil. Moreover, classes can oppose, influence, and dominate in many different ways. So multiple realizations look like a serious problem. That is doubly true if we are talking about how these processes are brought about by individual behavior. Here a fact frequently asserted by individualists – the creative and open-ended nature of human behavior – actually argues for the holist position. Classes dominate by multiple methods; those methods can be realized by diverse individual behavior.

We could make similar arguments for all the terms listed above. Most or all are defined in part in terms of their function vis-à-vis other social entities; most abstract from individual detail. It is just these features that make open-ended realizations in individual behavior likely, so we can expect most or all of Paige's other social predicates to be ineliminable as well.

The failures of individualism

Two other obstacles seem likely to thwart any reduction of Paige's theory. Paige does frequently refer to individual behavior. But he describes individuals in terms of their social roles – for example, as members of an elite class or as a certain kind of property holder. However, these individualist descriptions arguably presuppose explanations in terms of classes and judicial institutions. Consequently, they do not eliminate social explanations but presuppose them. Moreover, individualist descriptions of peasant political behavior are likely to be context sensitive. For example, acts of violence against upper-class individuals might or might not be counted as social rebellion. Paige argues, reasonably enough, that for peasant behavior to constitute a revolutionary movement it must (a) be outside existing institutions and (2) involve some sense of collective identity. Two identical acts of violence at the individual level might have different connotations at the social level. If some individual act can be part of a group with political demands in one instance and in another be an isolated or apolitical act, then that act has no unique social effect – and its effect depends on social context. As we saw earlier, this context sensitivity undercuts reduction. Since the multiple realizations problem looks overwhelming, I shall not pursue these other problems in any more detail.

Thus Paige's work certainly looks irreducible. So does that of Hannan and Freeman. And, we saw above that alleged cases supporting the reduction of social theories to individual ones do nothing of the kind. Of course individualist theory might be transformed and these problems be eliminated.[22] Yet from what we can see now, that prospect looks highly unlikely. These facts do not prove irreducibility for all time, but then probably nothing could. Nonetheless, they do provide, based on our best evidence to date, an overwhelming case against reducibility. Moreover, we have good reason to believe that our results hold across the social sciences. Rational choice theory is by far the most powerful and widespread individualist approach; the class and institutional analyses of Paige and Hannan and

---

[22] There are various ways the obstacles to reduction might be sidestepped by developments in social and individualist theories: (1) social kinds with multiple realizations could be fragmented into subkinds that had unique realizations in terms of individuals, (2) individualist theories might eventually reduce the social predicates they presuppose, (3) we might eliminate context sensitivity by reducing the relevant social context. Yet all these possibilities are merely that – and we have no evidence that such developments are likely. Presupposed social predicates and varying social contexts are likely to be multiply realized; they certainly are in the work discussed in this chapter. The natural basis for fragmenting social kinds would be by social traits. Thus these various possibilities seem no more than mere promises unsupported by any evidence. (For a careful sketch of these possibilities, see Causey 1977. Causey, however, provides only the possibilities, not a body of evidence in their behalf.)

Freeman are likewise instances of an approach widely employed in the social sciences. So our results should generalize. Furthermore, we have seen that reduction looks unlikely wherever we abstract from individual detail or locate social practices in a social web of causes. Most social science proceeds in exactly this way.

Let me conclude this section by considering one last objection, in the process drawing some positive morals about good social science. The ideal of a unified science motivates reduction. Individualism embodies that ideal by trying to show that the social science can be unified with sciences more general and fundamental. If we give up reductionism, do not we also give up the goal of a unified science? The goal of a unified science is a persuasive one. Yet rejecting individualism does not force us to give up that ideal, for reduction is only one way to get unity. A look at real scientific practice shows a more complex route to unity. As I have argued elsewhere (Kincaid 1990d; in press a), cell biology and biochemistry are integrated not by reduction but by other systematic connections. The cell biology–biochemistry case suggests a different picture of scientific unity, one that involves at least the following:

> showing that the entities of one theory exhaust those of the other
> showing that two theories make no inconsistent presuppositions
> showing that the predicates of one theory supervene on those of another
> developing extensive evidential interdependencies
> developing extensive heuristic interdependencies
> developing extensive explanatory interdependencies

This unity-without-reducibility model is, I think, generally a more accurate account of how scientific theories at different levels relate. It is also a model of scientific unity that holists in the social sciences can happily embrace, for it leaves the social with an essential place. Paige's work, for example, provides exactly these kinds of interdependencies between his account of large-scale social structure and his postulated mechanism in individual motivation. Individualist demands for reduction reflect a simplistic and outmoded picture of scientific unity.

5.2 Claims about explanation and confirmation

So far we have argued against individualism understood as a reductionist thesis. Individualists certainly do put their doctrine in such terms. Yet sometimes individualists try to skirt the reduction issue and claim instead that individualist theories can *explain* everything in the social world. Obviously they can do so only if this explanatory primacy can

The failures of individualism

be separated from reducibility. This section looks at that prospect and related individualist theses about explanation, in particular, the claim that individualist mechanisms are necessary to explain and that individualist theories are the best explanation. We also take up individualism as a doctrine about confirmation. Claims about the best explanation or about mechanisms are sometimes really claims about evidence or confirmation, as are the supposed heuristic values of individualism. Taken together, these theses largely exhaust the non-reductionist versions of individualism.

5.2.1 *Full explanation without reduction?*

Several different theses may be at work when individualists claim to explain without reduction. They may be making a claim about what is necessary or what is sufficient as well as about full or partial explanations. So at issue are at least these theses:

(E1) Any explanation of social phenomena must refer solely to individuals, their relations, dispositions, and the like.
(E2) Any fully adequate explanation of social phenomena must refer solely to individuals, their relations, dispositions, and the like.
(E3) Individualist theory suffices to fully explain social phenomena.

Thesis (E1) makes the most stringent claim, (E3) the least. Since (E1) and (E2) talk about what is necessary, they clearly presuppose (E3), which asserts only that individualist theories can explain. So if (E3) is implausible, then the other theses fall as well. However, there are good reasons to doubt that individualist theories suffice to fully explain.

Thesis (E3) faces the following dilemma: Individualist theory explains events as either kinds or tokens. An event "kind" (recessions, revolutions, and so on) has repeated instantiations; it does not refer to a specific event. A token is a particular event, one that is non-recurring and localized in time and space – for example, the French Revolution is a token event in this sense. Now if social phenomena are to be explained as types, thesis (E3) turns out to be nothing but an already rejected claim about theory reduction. Thus (E3) must therefore claim that social *tokens* can be fully explained individualistically. Individualist explanations of the latter sort, however, are certainly not fully adequate as the individualist asserts.[23]

Let me unravel the above argument by considering the claim that we can explain every social type individualistically. Explanation is done by theories, and the individualist is now claiming that individualist theory

---

[23] I develop this argument in much more detail in Kincaid (1988, in press a).

can in principle do all the work of social theory. Thus, there is allegedly for each kind of social event or entity an explanation in solely individualist terms. However, that means the individualist must supply an individualist equivalent for the types picked out by the social kind terms. Therefore, this explanatory version turns out to be, or at least require, reduction of theories – in short, the strict reductionist thesis. That thesis, however, has already been shown implausible.

The second alternative is to explain social phenomena in effect case by case. While we might not be able to give a single individualist account of social event kinds, we could still explain the social realm completely by explaining every particular social event. I can see two ways that explaining social tokens one by one might provide full explanations. One possibility is that social phenomena are fully explainable in that we can simply describe – for every particular social event – the individuals involved, their interrelations, and the like. Because social entities and events are realized in or supervenient upon individuals, we can describe that event by referring only to individuals. We can presumably name and describe the particular individuals involved and their interrelations in solely individual terms, in physical terms if need be. Of course, simply describing what individuals did to realize a given social process may be not much explanation at all, for it might neither cite causes nor unify.

A more interesting case is the following: There are individualist laws or causal generalizations describing human behavior that apply to the particular behaviors realizing a specific social event, even though the combined predicates of those laws are not coextensive with ones in social theory. In other words, the configuration of individuals as a whole might fall under a social description, while the individualist laws, for example, the laws of psychology, might apply to individuals one by one in a way that allowed no equivalence of social and individual terms. Such a situation would be analogous, for example, to computer programs that have potentially infinitely many realizations: we may not be able to define programs in terms of machine states, but there are still laws to explain any particular realization in physical terms. Thus we seem to have here a real possibility for individualist explanation.

These case-by-case explanations are likely to be incomplete for two reasons: (a) they may well presuppose social theories and (b) they will leave many questions unanswered. We saw the first problem earlier. A purely individualist account must make no essential use of social predicates or explanations. That means these case-by-case accounts must presuppose no background social structure. They cannot, for instance, describe individuals via their role in social institutions. Thus the individualist must employ very austere explanatory resources – or fail to explain solely in

The failures of individualism

individualist terms. As a result, individualist accounts are unlikely to give full explanations.

More important, explaining social phenomena case by case in individualist terms will not provide *complete* explanations, for important questions will be left unanswered. More specifically, for any social phenomenon or event S, two fundamental questions are:

(1) Why did S occur, that is, what are the causal connections between the kind of social events (events involving social entities) preceding S and the onset of S?
(2) Why do these kind of events (the kind to which S belongs) occur, that is, what other kinds of antecedent social events might bring about this kind of social event?

Question (1) could include questions about what kind of preceding changes in political, religious, intellectual, economic, class, or educational institutions causally influenced S. Paige, for example, asked about the causes of agrarian political movements. The causes included class structure, urban political parties, and the like – in short, *kinds* of *social* events. Question (2) seeks to place S in a general framework by citing the various possible causes of S. Hannan and Freeman, for example, give various routes by which specialist organizations might persist; those routes involve the various kinds of social environments present, namely, stable and fine-grained environments. Thus any theory of social phenomena that cannot answer these two kinds of questions is prima facie not a complete explanation.

Individualist explanations of social tokens will not be able to answer questions falling under (1) and (2) for a simple reason: they have no way of referring to kinds of social events. Both (1) and (2) involve situating a particular social event in a web of causal laws or generalizations connecting different social events. Causal generalizations, however, are about kinds of events, not event tokens. Thus the individualist who claims to give a full explanation – despite having no individualist terms coextensive with social kinds – must be in part wrong. Without tools to specify social event kinds, questions like (1) and (2) cannot be answered. Any theory that fails in this regard cannot claim to be full or complete.

We can see more clearly where the individualist goes wrong by noting a couple of points about explanation: explanation is often (a) contextual and (b) under a description. Following much recent work, we have considered explanation an answer to a question. As we saw before, questions are not answered *simpliciter*. Rather they are answered relative to a number of contextual parameters – for example, the contrast class and the relevant kind of answer. The idea that explanation is explanation under a description makes a related point. The same event can be described in multiple ways. Those different descriptions can then ground or call for different explanations, which may well be compatible.

These two facts about explanation make it clearer why individualist explanation case by case fails to fully explain. A complete account of what individuals did is an explanation under only one of many relevant descriptions. For example, the question "What caused the revolutionary movement in Vietnam?" may be answered individualistically for one set of contextual parameters – namely, those having to do with microstructure. However, if our relevant contextual parameters are social, then the individualist account will not do. The question "Why did a revolutionary peasant movement occur (rather than a reform movement or a commodity reform movement or . . .)?" demands an answer in terms of *kinds* of social causes.

So we can thus see why supervenience does not ensure that the individualists have a full explanation, contra its defenders.[24] Supervenience may entail that we can describe what individuals did, for example, in bringing about the French Revolution. We might go on to invoke laws of psychology or other laws about individuals to say why they behaved as they did. Such a story would be explanatory, but it surely would fall far short of being a full explanation. We would have no way to understand this event as a social kind. Consequently, we could not explain its causal connection to preceding changes in classes, religious institutions, and the like, nor could we understand this revolution by relating it to other revolutions, other kinds of revolutions, or political and economic transformations that have or might occur elsewhere. In short, many standard questions are left unanswered.[25]

So far I have argued that individualism cannot claim to fully explain without reduction. That argument rests on two crucial assumptions: (a)

---

[24] As has been claimed by McDonald and Petit (1981, p. 125), in an otherwise excellent book.

[25] Miller (1978) argues that individualist accounts also cannot answer questions about why social events had to happen. Describing the actions of Robespierre and his fellow revolutionaries will not explain why the French Revolution was inevitable, even if Robespierre had died before the revolution began. I am sympathetic to Miller's point, but it is unclear to me (1) how much social explanation really can make a claim to show inevitability and (2) that a *full* account of all the individuals involved would not suffice to ground claims about inevitability. For example, it is obviously controversial to claim that the French Revolution was inevitable even if its main participants had acted differently, and the individualist can make sense of the claim that the revolution would have happened without Robespierre by referring to the causal importance of other individuals.

that there can be or there are well-confirmed *holistic* social explanations invoking kinds or types, and (b) that questions about macrosociological processes have to be answered for a full explanation. If all theory referring to social entities is a failure and is doomed to be so, we cannot fault purely individualist explanations for not answering questions about social kinds. Furthermore, any scientific theory, no matter how good, will probably leave some questions unanswered. So why think that individualist explanations are incomplete because they cannot answer questions posed by social theories?

The thesis that holist social theories are inherently inadequate simply because they are *holistic* (thesis [E1]) is at first glance highly implausible. We have already argued at length in Chapters 3 and 4 that social theory can and sometimes does produce well-confirmed causal explanations. We have also argued that our examples of good social theories are also irreducible. Thus we have seemingly provided strong counterexamples to the thesis that holist theories cannot explain. Furthermore, we shall also argue in the next section that there is no inherent obstacle to holistic *evidence*.

I can see two reasons why someone might think that any non-reduced social theory is inherently non-explanatory: one based on the idea that theories explain only if they can be unified with the rest of science, and another based on doubts about macrosociological causation. Neither rationale carries much force.

Early advocates of individualism thought that the unity of science required reduction. A unified science was an important goal to the positivists in part because they thought it essential to *explanation*. By reducing the higher-level sciences to their more fundamental and general counterparts, we increased their claim to explain. However, so the reasoning goes, unreduced holist social science forgoes such ties to more basic theories and thus is explanatorily suspect.

There are three reasons why this argument is implausible:

(1) It is motivated by a picture of explanation we have already criticized – the idea that explanation is essentially unification. If unification is the key to explanation, then irreducible social theories might seem explanatorily suspect because they seemingly give social explanations no place in a unified overall picture of the world. But this demand that good explanations must be embedded in other, more unifying accounts carries no automatic weight if explanation is essentially about causation. The world may be a causally complex place. Causal processes at one level of description may not be easily subsumed under causal explanations at other levels of description.

(2) There is compelling evidence that much good natural science cannot be embedded in more fundamental theories via reduction. The claims of evolutionary biology are about a basic predicate, fitness, that can be brought about by indefinitely many different physical traits. Similarly, I have argued elsewhere (Kincaid 1990d; in press a) that reduction is unlikely even in *molecular* biology, precisely for the same kinds of reasons that thwart the reduction of social theory. So if holist social explanations are inadequate, so is much natural science. The reasonable conclusion is surely that the philosophers' stricture on explanation is inadequate.

(3) Holist theories *can* have non-reductionist ties to more fundamental theories. We can see how large-scale social processes are brought about case by case. We can sometimes find typical mechanisms undergirding macrosociological processes. While I shall argue later that such mechanisms are not essential to explanation, they nonetheless are sometimes possible and do help provide a more complete account.

So macrosociological explanations are not inherently flawed as explanations simply because we cannot tie them *via reduction* to more fundamental and general theories.

Worries about macrosociological causation might also ground doubts that holist theories explain, for causal relations between social entities may seem suspect.[26] I think these suspicions are misplaced. There are three ways we can read macrosociological causal claims: as claims about (a) complex wholes, (b) logical aggregates, and (c) non-logical aggregates. Macrosociological causation can be coherent on any of these readings. Consider first macrosociological claims about complex wholes. A complex whole is a unitary entity with internal structure and complexity. Many causally efficacious social entities are complex wholes. Labor organizations, bureaucracies, some political parties, corporations, non-profit organizations, and so on are arguably complex wholes. However, complex wholes can surely stand in causal relations. If they cannot, then the only real causal relations are at the level of fundamental particles, since the natural sciences are largely about complex wholes – planets, organisms, and the like. Thus if the natural sciences explain, then macrosociological explanations invoking complex entities are not flawed simply because of their aggregate nature.

Consider next the case of logical aggregates. By a logical aggregate, I mean a sum that is simply constructed – the components for all intents

---

[26] Randall Collins (1988, p. 393) makes this claim.

and purposes stand in no real causal relations to one another. Nonetheless, such aggregates can also stand in causal relations. Causal claims are generally "summative." If A_1 causes B_1, A_2 causes B_2, and so on, then the totality of A's causes the totality of B's. To see this, we can plug in these causal claims into various theories of causality. For example, the counterfactual "if the totality of A's had not occurred, the totality of B's would not have occurred" is true. Similarly, if there is an invariable succession between each A and B, then if all the A's occur together, then so will the B's. In short, an invariable succession will hold among the totalities. So logical aggregates can be real causal factors. This should not be surprising, since causal claims about them are just shorthand for claims about component causes.

Finally, consider non-logical aggregates. Here I have in mind aggregates whose parts do interact but which lack the kind of internal structure we usually require for complex wholes. Some aggregates along these lines might include aggregate income, total unemployment, or aggregate criminal activity. These aggregates can also stand in causal relations. Again, the best way to see this is simply to plug such aggregates into various accounts of causality and notice that nothing in principle prevents true causal statements from resulting. Of course, there are *epistemological* difficulties, difficulties we will take up later. In particular, we must always wonder whether the aggregate we have identified contains too much, as it were – if the true cause involves some lesser part of the aggregate (though note that when I claim the baseball broke the window, I seldom know that some lesser part of it would not have sufficed). But if we have evidence, for example, that if aggregate A had not been present, then aggregate B would not have been either, then we have reason to believe A caused B at least on one theory of causation.

We have thus found no convincing case that holist social theories, simply because they are holist, cannot explain. What about our second assumption that social theories answer questions that any full account must answer? Science sometimes does progress by eliminating or ignoring questions. However, it usually does so because those questions do not lead to fruitful answers. We saw that no such case can be made about all social explanation. Moreover, we have both strong theoretical and pragmatic reasons to think explanations in social terms are important. Let me mention three:

(1) I argued earlier that individualist explanations frequently presuppose holistic ones. That was true for rational choice theory and Homans's small group theory; it was true for the individualist parts of Paige's account. When this happens, the individualist is

willy-nilly committed to taking social explanations seriously – or to eschew explanation altogether.

(2) Individualist theory may need holistic accounts in many other ways: social kinds may identify patterns not easily seen at the individual level, may suggest fruitful questions, and may play a key role in confirmation. In short, those who want to develop a theory of individuals may have strong heuristic and evidential reasons for taking social explanations seriously.

(3) Finally, we have strong moral and political reasons for wanting social questions – that is, questions framed in terms of social entities – answered. What forms of political organization are just? most stable? What economic arrangements are just? stable and efficient? What explains the prosperity of some social groups and the poverty of others? The list of fundamental social questions goes on and on. *If* reductionism were true, then individualists could claim to answer these questions. Without reduction, they cannot.

In the end, simply denying that social questions need be answered is hardly plausible – it is winning the debate by changing the subject.[27]

Thus we have removed the last objection to our argument that individualist explanation without reduction is seriously incomplete. Without the aid of social kinds and the causal generalizations they allow, case-by-case individualist explanations will leave many questions unanswered. Avoiding these problems by explaining social *kinds* brings us back to reduction. But we have already argued at length that reduction is improbable. So the attempt to drive a wedge between reduction and explanation fails.

Before we move on to other claims about explanation and evidence, I should note that the results of this section make good on a promise made earlier. In discussing reduction I pointed out that actual reductions fall on a continuum running from pure reduction at one end to elimination of failed theories on the other; my arguments against reduction were directed at reduction in the pure sense, though I promised to take up eliminativist versions later. Now we have an answer to those who defend reduction by elimination, in other words, an answer to those who claim that holist theories are reducible because only individualist theories can succeed. The arguments sketched so far show (a) that there is nothing inherently wrong with macrosociological theories, (b) that holist theories pursue important questions, (c) that some holist theories are relatively well confirmed and

---

[27] Here I disagree with Toumela (1990) in his discussion of my previous work on individualism. See Kincaid (1990b) for a more detailed reply.

explanatory by the standards of good natural science, and (d) that individualist theories do not fully explain. An eliminativist route to reduction is thus highly implausible. We shall give further evidence for that conclusion as we turn next to look at the idea that individualist theories are the best explanation.

5.2.2 Is individualism the best explanation?

Another common individualist thesis is that individualist theories are *best*-confirmed or *most* explanatory.[28] Individualists frequently claim that individual variables explain some phenomenon, while holists argue that social variables must at least be included. This way of putting the issue is somewhat orthogonal to other theses I have listed – and it easily generates confusion. Moreover, what clear theses we can derive from this formulation of individualism are quite implausible.

Before we can evaluate this version of individualism, we need to clarify the notion of "best explanation." It is quite common to switch interchangeably between the best explanation and the best-confirmed account. That identification is not surprising, for when the idea of "best explanation" is examined, frequently it involves no more than predictive scope, coherence with background information, and the like.[29] However, "best explanation" in this guise is really nothing more than overall empirical adequacy, and the best explanation and the best-confirmed theory are one and the same. If we want the best explanation to be something different from overall empirical strength, then we need to make substantive assumptions about what explanation requires. For example, if citing causes is essential to explanation, then we could judge theory A to be a better explanation than theory B if, say, A provided causal laws and B provided only behavioral laws. Thus "best explanation" can take on a separate content if we presuppose some substantive account of what explanation involves. Note, however, that while "best explanation" is given a real independent content in this way, the best explanation in this sense need not always be the best theory. Explanatory success in this sense is only one of multiple virtues; a theory that used more causal laws, for example, could do less well than one that invoked more behavioral laws when it came to other empirical virtues such as predictive scope and so on.[30] Individualists and holists do not clearly separate these different senses of "best explanation," but we need to do so to evaluate their claims.

[28] This claim is made by Toumela (1990) and Collins (1988, p. 390), and it is not an uncommon way of putting the issue.
[29] Gilbert Harman (1965) glosses "best explanation" in this way.
[30] Here I am summarizing conclusions reached elsewhere. See Day and Kincaid (1994).

There is a second confusion we need to dispel. To say that individualist theories explain better or are better confirmed assumes that they and holist theories *compete*. Yet holist theories and individualist accounts are not necessarily competitors. They are theories at different levels and thus may be two different descriptions of the same reality. For example, Kepler's laws and those of quantum mechanics cite different variables, yet they need not be incompatible, for they can apply to the same objects differently described at the same time. Something similar might hold for individualist and holist theories. Thus claiming that one is a better explanation would make little sense.

What does it take then to show that holist and individualist theories compete and thus that one can be the better explanation? Translating holist explanations to individualist ones would be the ideal way to make this decision, but that route is not available – we are looking for plausible individualist theses that are independent of reducibility. Thus we must look for less systematic connections between theories at different levels that help us determine if they are incompatible. Those connections will have to begin either from above or from below. From above, we could (a) look one by one at how individual behavior realizes social variables in terms of individual behavior or (b) determine what sorts of individual mechanisms those social variables presupposed. From below, we could try to determine case by case how individual variables aggregate. Holist and individualist theories will only compete if aggregating or disaggregating produces explanations that the other theory cannot accept.

Only after we show that the holist and individualist theories compete does it make sense to claim that one is better. So individualism as a claim about best explanation must show (1) that all well-confirmed social theories are competitors to individualist accounts and (2) that individualist theories are always best-confirmed or the best explanation.

Let me illustrate these conditions with a realistic example. Sociologists and neo-classical economists currently propose different explanations for the distribution of income to individuals. On the neo-classical view, labor's income is determined by, among other things, its marginal product. That in turn is influenced by inherent skill, investment in education ("human capital"), and other individual characteristics. Sociologists claim that those factors are minor relative to social causes such as bargaining power or economic norms. It might seem that these explanations compete, yet that may not be so. If we ask what individual processes realize social power and norms, we may identify variables that are not at odds with those of the individualist neo-classical approach. Power and norms might simply be cashed out as facts about preferences, prior distribution of wealth, relative scarcity, and other factors that underlie supply-and-demand curves.

But the neo-classical account does not deny that supply and demand are a factor. Thus to show that the two theories are real competitors, we would need to show at least the following: (1) that the sociological approach made claims about individual processes incompatible with the individualist processes identified by the neo-classical account (or vice versa) and (2) that the neo-classical individualist processes had the best evidence or were most compatible with reasonable constraints on explanation. Only then would the two theories be real competitors and only then would it make sense to claim that one was a *better* explanation than the other.

There is little reason to think that the holist theories defended earlier necessarily compete with individualist accounts. Nor is there good reason to think that promising individualist theories such as rational choice accounts must be incompatible with social explanations. Recall that Paige's theory of agrarian political behavior is a *mixed* theory: it explains in terms of both macrosociological processes and individualist mechanisms. Because it integrates its holistic explanations with individual-level processes, there is a strong prima facie case for thinking it will not compete with a good theory of individual behavior. Similarly, Hannan and Freeman's macrosociological account abstracts from the internal structure of organizations. Thus they presuppose no very specific set of individual behaviors and motivations; those are left for organizational psychology to fill in. However, because Hannan and Freeman's theory is compatible with a broad range of underlying behaviors, it is likely to be compatible with whatever is the best individualist account. Furthermore, the individualist theories we examined earlier – those of Homans and the rational choice school – all presupposed social explanations. If they presuppose social explanations, however, they cannot be incompatible with all macrosociological accounts, on pain of inconsistency. Thus the best cases of individualism probably do not compete with social explanations.

Moreover, we can make a case that when social and individualist theories do compete, social theories will sometimes win out. If we take "best explanation" as "best overall confirmation," the macrosociological accounts may have predictive scope on their side. Sans reduction, individualism cannot capture causal patterns at the macrosociological level; what looks diverse from the viewpoint of individual behavior may constitute a kind at the social level. At least in this sense, holist theories ought to have more predictive scope.

This fact explains, I think, why holists are tempted to put the issue between them and the individualists as one about the best explanation. Individualist theories often are of the radical variety – they pick some simple individual traits, often monadic traits, and claim they explain large-scale social behavior. Explaining poverty as the result of some specific individual personality trait is an obvious example. Holists are right to be suspicious and to think they offer better accounts in such cases. But it does not follow that their accounts are necessarily better than or competing with all individualist explanations, no matter how subtle. In fact, since social events are brought about by individuals, social explanations must be compatible with at least some account of what individuals do.

Holist theories may also do well if we spell out "best explanation" in terms of some specific account of explanation. If explanation is citing causes, then holist accounts can identify causal processes that lower-level accounts cannot, namely, those described at levels above the individual. More important, as we shall discuss in the next section, holist theories need not eschew providing mechanisms, including mechanisms involving individuals. (Recall that the holism defended here only claims that macrosociological explanation has an essential role, not that it is all there is to social explanation.) Thus holist accounts can invoke both macrosociological causal patterns and underlying causal detail, something the individualist cannot. So the holist account would seem to do a better job of citing "the causes," if that latter notion makes sense.

In sum, holist and individualist theories often do not compete. When they do, holist theories may well be both better-confirmed and better explanations.

Before we leave this topic, we should note that logically compatible theories might differ as explanations for pragmatic reasons. One theory may make causal relations easier to see or express, may better show us which factors we can manipulate, or may have the resources to describe causes at the right level to answer our questions. While these factors are important for practical purposes, they do not make it essential to ask which theory explains best *unless* explanation is entirely a pragmatic affair. I do not accept such a picture of explanation. Causal explanation is an objective matter once the relevant questions have been set. Moreover, it is not at all obvious that individualist theories would fare well in pragmatic terms – the ability of social theories to abstract from messy individual detail, for example, may make them easier to use and so on.[31]

---

[31] If we take explanation to be unification, two compatible theories might still compete in that one unified more than the other. Yet would it be clear that a successful individualist theory provided more unity? If the holist theory is irreducible because of multiple realizations, then it finds patterns that the individualist theory does not. So it is not at all obvious that the individualist theory would be the better explanation in this sense. Of course the individualist theory might have other unification virtues. This, however, points out – as I argued in Chapter 3 – just how hard it is to use unification as any kind of objective criterion for good explanation.

5.2.3 *Are individualist* mechanisms *necessary?*

I want to turn now to the idea that individualist mechanisms are *necessary* for explanation or confirmation. Elster claims that no causal explanation involving social entities can be adequate until a mechanism is identified: "To cite the cause is not enough: the causal mechanism must also be provided, or at least suggested" (1989, p. 4). I take this claim in two ways – as a requirement for explanation and as one for confirmation – because Elster himself is unclear. However, his arguments suggest that his main concern is confirmation, though others such as Little (1989, ch. 7; 1991) have explicitly made the requirement one for explanation. For charity's sake, I consider both versions.

Elster claims we need to identify individualist mechanisms to confirm causal relations between social variables. I doubt that this is general truth, though there may be specific circumstances where mechanisms are needed. A first problem is that "the mechanism" has no definite sense. For any causal claim about aggregate variables, there are indefinitely many mechanisms. For any mechanism, M, that we cite between sociological variables A and B, we can go on to ask again how M makes the connection between A and B possible. We can postulate both "horizontal" mechanisms – those between M and A or B – and "vertical" mechanisms, that is, the lower-level processes or structures realizing A and determining its causal capacities. In either case, we can go on asking about "the" mechanism as long as we have not reached the fundamental level of nature.

Thus demanding that we cite the mechanism for macrosociological causation is inherently ambiguous. Do we need it at the small-group level or the individual level? If the latter, why stop there? We can, for example, always ask what mechanism brings about *individual* behavior. So we are off to find neurological mechanisms, then biochemical, and so on. Unspecified, "the mechanism" is of dubious sense.

Maybe Elster wants all the detail. If so, almost no causal claims are verified, for we almost never know the complete mechanism, whatever that means. Moreover, requiring all the detail would not just rule out holist theory – it would leave *individualist* accounts unconfirmed as well. After all, we know little about the neurophysiology underlying human action.

Elster's demand for the mechanism leads to other absurdities. He claims we need mechanisms to avoid spurious causation. Elster is right that no causal claim is confirmed until we show that connection is not the spurious result of some third variable causing both. However, we do not need to identify the precise mechanism to eliminate spurious causation. What we do need is some plausible list of possible mechanisms or, more generally, possible causal factors that might make the alleged relation spurious. Good statistical practice, as we discussed in Chapter 3, shows a correlation stable: it shows that an observed correlation holds up when other possible interfering variables are added in. If we take an apparently significant regression equation and run the equation again with any possible complicating variable, we can usually rule out spuriousness. If some third variable is responsible for the effect, then the correlation will disappear after we add in the true cause. This, of course, is elementary statistics. However, it suffices to show we do not need the precise mechanism, only some plausible list of possible mechanisms. This is a fortunate conclusion. If we always have to have mechanisms, very few causal claims would be confirmed, inside science and out.[32]

We can confirm causal claims without mechanisms. Nonetheless, mechanisms can be essential in those specific contexts where precise mechanisms are presupposed. We saw one such context in Chapter 4 in discussing functionalism. Functional claims sometimes use optimality arguments for evidence; optimality arguments presuppose that some mechanism exists bringing about the optimal. So in this case we need to know there *is* a mechanism, though we still do not necessarily need to know *what* it is.

When, then, do we need mechanisms to confirm macrosociological claims? I suggest three factors are decisive: (1) how confident we are about our evidence, assumptions, and theory at the macrosociological level; (2) how confident we are about our evidence, assumptions, and theory at the level of the mechanism; and (3) what exactly our macrosociological theory presupposes about mechanisms. When our macrosociological claim is weakly confirmed, when it makes quite specific assumptions about individual behavior, and when we have a well-confirmed account of individual behavior, then the demand for individualist mechanisms makes sense.

However, we are not always in this situation. A macrosociological claim might be well confirmed at its own level, it might make no very specific assumptions about individual behavior, and there might be no very well-developed individualist theory. Hannan and Freeman's work in organizational ecology is a case in point. They *do* cite a mechanism, but not an individualist one. Rather they identify macrosociological mechanisms grounding their claim. Since their hypotheses are tested with care and since there is no very well-confirmed theory of individuals in organizations, we should not require them to produce an individualistic mechanism. In fact, we can reasonably demand quite the opposite – that any

---

[32] Evolutionary biology is a prime case here. Darwin had good evidence for natural selection even though he had the mechanism partly wrong. Current work may frequently know that selection on a character is taking place without knowing how it is – at least any detail. See Endler (1986).

theory of individual behavior in organizations be compatible with known macrolevel processes. In short, Elster's best arguments for methodological individualism sometimes support a kind of methodological holism! Sometimes we must show the macrolevel relations before we can reasonably confirm a hypothesis about microlevel detail.

Let's explore one last possible motivation for Elster's stricture: that individualist mechanisms are necessary to *improve the precision* of macrosociological predictions. The rationale here is that macrosociological accounts work in terms of aggregates. To improve the predictive precision and scope of theories working with aggregate variables, the argument goes, we must reduce them to individualist terms or at the very least provide individualist mechanisms. Only in this way can we eliminate the complexities that limit predictive improvement.

As a requirement for predictive improvement, the demand for individual-level mechanisms fares no better. We can refine and improve macrosociological explanations in macrosociological terms by bringing in other macrosociological variables. Paige, for example, did exactly this when he added in the effect of urban political parties to improve and test his basic model. So did Hannan and Freeman when they added other populations of organizations to their basic model. This basic strategy – refining a theory by adding variables of the same type or level – is widespread in science. So there is no reasonable argument that micromechanisms, in particular individualist mechanisms, are *necessary* for predictive improvement.

Should we at least grant that individualist mechanisms are one important way to predictive success? Yes and no. Surely the natural sciences sometimes improve their accuracy by identifying lower-level mechanisms, and that lends credence to a similar hypothesis about the social sciences. Yet we have reason to be a bit skeptical here as well. Because macrosociological processes may be brought about by sundry types of individual behavior, lower-level detail may be too diverse and heterogeneous to tell us anything useful. Mechanisms are no universal requirement on confirmation; their actual role will depend on substantive, contextual facts that will have to be argued for case by case.

Elster's demand for mechanisms as a requirement for *explanation* is likewise implausible. If no causal account explains without the mechanism, then almost no causal claims explain. After all, we constantly give causal explanations inside science and out while never citing the relevant mechanisms. And, as before, "the mechanism" is of questionable sense, since mechanisms occur at multiple levels and so on. Do I have to cite the subatomic details to explain that the errant golf ball caused the broken window? If we really took Elster's demand seriously, then individualist accounts would not explain either, since they do not describe the neurophysiological mechanisms.

In sum, no universal demand for mechanisms is defensible nor would it support individualism.[33] Yet Elster's heart is in the right place, for mechanisms do play a role in some contexts, at least when it comes to confirmation. When optimality arguments are involved, when we have very well-developed theories at the lower level, then looking for lower-level detail can be very important – though "the mechanism" is not needed nor is a mechanism in terms of individuals. I argued in Chapter 2 that methodological injunctions are empirical claims dependent on background information and thus likely to be context dependent. The demand for mechanisms is no exception.

5.2.4 *Individualist evidence and heuristics*

We have two last theses about confirmation and explanation to consider: the idea that evidence about individuals is all the evidence and the idea that pursuing individualist theories is the best heuristic.[34] We can dismiss the former claim with some confidence; evaluating the latter one is more complex and our conclusion will be somewhat tentative.

We should ask first why all social evidence must be about individuals. This claim seems to rest on a general principle that a theory can be confirmed only by evidence about the components of the entities in its domain. That principle is highly implausible, for evidence about complex wholes like planets, organisms, and so on is obviously essential to the natural sciences. I suspect the demand for evidence about individuals is a bastardized version of an old positivist claim: the claim that theoretical terms must be defined in observational ones, in particular individual sensory experiences. That claim was jettisoned long ago for reasons I won't rehearse here.

Perhaps there are special facts about the social sciences that make individual evidence paramount. That seems unlikely. We have seen in Chapters

---

[33] W. Salmon (1984) sometimes claims that "mechanisms" or "underlying mechanisms" are necessary to explanation on his account. While I admire his account of explanation, I see nothing in it that requires mechanisms in the sense discussed here. Rather, what Salmon seems to call for is explanation via citing *causal processes*. Causal processes and the underlying mechanism, whatever that means, are not the same. Behavioral laws would on Salmon's account need causal mechanisms in the sense that we need to cite the causal process bringing about the purely behavioral law. However, this demand for mechanisms is one I share and is one quite different from that announced by Elster and others.

[34] The claim about evidence is an old one; it is made recently by Collins (1988, p. 391).

The failures of individualism

3 and 4 that there can be good science without evidence about individuals. Hannan and Freeman had data about the founding and longevity of *organizations* in an environment characterized in organizational terms. Paige had evidence about social classes, political parties, and the political behavior of groups. Hardly a fact about particular individuals appears in their work. Yet Chapters 3 and 4 made a case that these theories are relatively well confirmed.

Nonetheless, the claim that only individual evidence will do may have at least one plausible reading. Social science evidence is frequently evidence about *aggregates*. Aggregate evidence, however, raises the kind of worries that bother Elster – namely, that spurious relations are introduced and real relations lost in the process of aggregating. So it is worthwhile to look for a moment at this more specific doubt about social science evidence.

There are two senses in which the social sciences use "aggregate" evidence: either they may pool together information about each individual object under study to infer about causes acting on those entities or they may have evidence only about totals or aggregates. Data about the income of every criminal in a given population are of the former sort; data about the total income level and crime rate are of the latter sort. No special problems arise when macrosociological claims are based on data about individual social entities. If I have a batch of data about individual corporations, their profitability, and their corporate structure, then inferring causes here is in principle no different from inferring causes with similar data about individuals who smoke and have cancer.

However, when my data points are only aggregate values, then causal inference can be more troublesome. Using correlations between aggregate entities like total income and total education to infer about the causes of individual behavior can lead to mistakes. The correlation between aggregates may be a by-product of how the data are aggregated. Census information, for example, gives us purely aggregate information where aggregation is by geographical areas. If the geographical area is sufficiently large, it may combine rich and poor, educated and uneducated into one unit. Data about such units may thus show little correlation between education and income, since the factors cancel out because of how they are grouped. In short, having evidence only about aggregates does not tell us how the relevant variables are distributed over individuals. Yet we need that information to make inferences about the units we aggregated.

This problem with purely aggregate evidence is serious. However, it is no fatal flaw to the holism defended here for two reasons. First, there are ways to minimize the problems. After all, aggregate evidence is not unique to holistic theories. In effect, we need to control for whatever possible factors may be introduced by aggregation. For example, finding homogeneous aggregates along the relevant dimensions helps control for spuriousness. In the education and income example cited above, that would mean picking geographical units that we knew had little variation in income. Social scientists have identified numerous other methods for dealing with aggregation bias (Langbein and Lichtman 1978). With the right background information, aggregate evidence can be reliable. Furthermore, these aggregation problems do not arise when using aggregate evidence to argue that one aggregate causes another, for we are making no direct inference about individuals.

Aggregate evidence is not a serious problem for the holistic research cited in earlier chapters. Hannan and Freeman's work does not rely on correlations between aggregate data. Their evidence is about individual organizations and their traits. So they do not run into the special problems that purely aggregate evidence brings. Paige does use aggregate evidence, but he is aware of the dangers and thus constructs his aggregates so that they are homogeneous for the causal variables under study. Paige also supplements his statistical causal model with detailed case study evidence. While these efforts do not ensure that aggregation causes no problems, they do make it unlikely.

So we can safely reject the assertion that all evidence must be about individuals. A related and perhaps more plausible individualist thesis is that pursuing individualist theories is the best *heuristic*. A heuristic, as I am using the term, is a research strategy. For example, "apply differential calculus," "start with simple models and add complications," and "approach every question about cell functioning as a problem about gene expression" are widely used heuristics in the natural sciences. Heuristics in this sense are useful research strategies, not foolproof or essential recipes for success. This distinguishes heuristics from other individualist methodological constraints which make mechanisms, reduction, and so on necessary conditions for good science.

Taking the approach outlined in Chapter 2, a good heuristic is one that the empirical evidence shows promotes good science, in particular well-confirmed causal explanations. Making such judgments about heuristics, however, runs into numerous complications:

(1) The "individualist heuristic" is really many different heuristics, depending on the version of individualism at issue. For example, "seek reductions" is a different recommendation than "seek mechanisms," and the success or failure of one need not entail the same about the other.
(2) "Well-confirmed causal explanations" are scientific goals described at their most abstract; in practice, that goal is embodied

The failures of individualism 185

in much more specific goals which specify the kind of causal information, the kind of evidence, and so on, that is necessary. This makes evaluating heuristics complicated.

(3) "What the empirical evidence shows" will not at all be obvious. The problems here go beyond the obvious difficulties of proving cause and effect in something as complex as scientific practice. There are special difficulties associated with generalizing even when we know causes in a particular scientific context. A heuristic that is fruitful in one situation – for example, in one historical period or one discipline – need not be so in another if scientists have importantly different views about the world and if the social structure of science is markedly different.

So arguing for or against heuristics is a difficult business. That said, I think there is good reason to be weary of individualist strategies, at least in their most virulent form.

"Seek reductions" and "seek purely individualist explanations" are likely to be bad advice, given the obstacles to reduction outlined earlier. Reduction fails because social predicates can be brought about in multiple ways, because individualist descriptions do not fix what happens in the social realm, and because explanations of individual behavior presuppose an institutional context. These possibilities make a reductionistic heuristic questionable. For example, when social phenomena have multiple realizations, looking solely for individualistic mechanisms will lead us to see diversity where there really is unity. In other words, concentrating solely on individual processes will miss patterns at the social level. On the other hand, looking only at individual facts will lead us to see unity where there is none. Social predicates that are context sensitive will lead to false generalizations. For example, imagine that a particular sort of consumer behavior is associated with macroeconomic recession. If that tie is context sensitive, then assuming that this consumer behavior leads to recession elsewhere is an error – for the context may be different.

Ignoring institutional detail presents different problems. It means important variables will be left out. If we look at the most successful social science research in the individualist tradition – for example, rational choice approaches – we will see that they do not explain solely in individualist terms. So those successes, if they are such, do not argue for strong individualist heuristics. Rather, they suggest that proceeding in purely individualist terms will be a kind of specification error and will make it harder to confirm even individualist claims.

The history of science may tell us something about individualist heuristics. Yet squeezing out the message is no easy task. Surely reductionist approaches sometimes promote progress, but they do not have a spotless record. During the nineteenth century, vitalism dominated biology. Vitalism is the idea that a non-physical force drives biological organisms. This antireductionist doctrine apparently led to scientific progress. Reductionistic approaches at the time were so crude they had little to offer. Vitalism justified pursuing biological phenomena in their own right, allowing reductionism to take its rightful place later.[35] Other parts of biology have similar lessons to teach. I have argued elsewhere that cell biology is not reducible to biochemistry and that pursuing purely biochemical approaches may not promote scientific progress. Reductionistic strategies have also caused problems in pursuing the theory of evolution (Wimsatt 1980). So those like Elster (1983, p. 53) who appeal to biology to support individualism are probably on shaky ground.

What about weaker individualist strictures – for example, "seek individualist mechanisms"? We argued earlier that mechanisms are not *essential* to confirmation or explanation, and that their usefulness depends on contextual background information. That general moral applies to seeking mechanisms as an individualist heuristic. Whether seeking individualist mechanisms produces greater progress will depend on questions like the following:

> Do we have a well-developed and well-confirmed theory at the individualist level?
> Do we have good evidence that the likely mechanisms are few in number and relatively universal?
> Does our macrolevel theory make strong assumptions about what individual-level processes must be like?
> Is our macrolevel theory poorly supported by the evidence?
> Is it difficult to refine it by adding further complications at the macrolevel?

Affirmative answers will bode well for the individualist heuristic, negative answers will argue against their adoption.

Answering these questions is an empirical matter that must be done case by case, and I cannot claim to provide compelling evidence here. Yet the arguments of this chapter do give us reasoned opinions about what the evidence is likely to show. We have seen that social processes are likely to be realized in diverse individual-level detail, that macrolevel social theories can be well confirmed and refined at their own level, and that macrolevel social accounts can proceed without any very strong assumptions about

---

[35] See Lenoir (1982) for a discussion of the various roles that vitalism and teleology played in nineteenth century biology.

individual behavior. Hannan and Freeman's work is an obvious case in point. So at the very least even the weak individualist heuristics are likely to be ill advised for an important range of social research.

Doesn't the success of modern economics and the rational choice paradigm show that seeking individualist mechanisms is at least *often* the best heuristic? Maybe so, but the answer is not as obvious as it may seem at first glance. Modern neo-classical economics has been successful. Yet its successes are arguably much more concerned with developing elegant mathematical models than with providing well-confirmed causal explanations, as we will point out in Chapter 7. Moreover, we will see that where economics is successful in providing well-confirmed causal explanations – in supply-and-demand analysis of actual markets – it does so not by seeking individualist mechanisms but by working at the macrolevel.

Finally, even when seeking individualist mechanisms is a useful research strategy, we need to keep in mind just how little that shows. For example, suppose that seeking rational choice explanations was sometimes the best route to success. Given the arguments earlier in the chapter, seeking those explanations will end up producing social theories that are through-and-through social. In general, it is possible that heuristics of one kind will produce theories of another. For example, "treat gravitational forces as a two-body problem" is a heuristic that has done much to support a theory that says just the opposite. So even when seeking individualist mechanisms is fruitful, it may well offer little support to the overall individualist program.

5.3 A question of ontology?

Despite the many different individualist assertions attacked in previous sections, we have ignored individualism as an ontological thesis. Yet the individualism–holism debate would seem at first glance to be a debate in ontology. Individualists are ontological conservatives, cringing at the promiscuousness of holists whose world is populated with collective mentalities, group beliefs, and supraindividual forces. In this section I use the results of previous sections to say something about the ontological issues. My discussion will be brief for several reasons. One standard approach to ontological commitment – Quine's – makes it a matter of what our best theories refer to. So our previous discussions will have already largely argued against individualism on that criterion. However, if Quine's criterion is inadequate (contra my own sympathies), then resolving the ontological issues would lead us into metaphysical disputes far removed from live issues in social science research, my topic of interest. Thus I will not engage these more traditional metaphysical disputes in detail.

The most obvious individualist thesis about ontology is that only individuals are real. We have already provided much evidence to the contrary. Social theories are ineliminable; our best explanations of social phenomena quantify over (refer to) social entities such as classes, government institutions, organizations, and the like. We are committed to their existence. Moreover, we have argued that such entities can stand in causal relations, both to other entities and to individuals. So they are "causally real" as well.

Individualists have always worried that making social entities real means denying some fundamental truths about individuals. The truths they have in mind are the ones Watkins called "metaphysical commonplaces," namely, that society does not exist and act independently of individuals. This latter claim we spelled out as supervenience. The former claim – that society does not exist over and above the totality of individuals – is what physicalists have called an "exhaustion" principle. Individuals exhaust the social world in that every social entity is composed of them.

These principles are indeed plausible, though they are not *a priori* truths. Indeed, important empirical issues are lurking here. If the social facts supervene on the individual facts, it is an open question which individual facts are relevant. Those facts might be biological, psychological, and even social, for class or institutional position could in principle be involved as well. So many of the debates sketched in previous sections will resurface in spelling out Watkins's commonplace. The exhaustion principle also hides similar complexities. Aside from individuals, society might be composed of material goods as well. How we describe those goods may presuppose answers to complex issues in the social sciences, particularly economics.[36]

However we formulate these "obvious" truths about individuals, we can still consistently allow social entities into our ontology. We can identify social entities with sums of individuals; individual behavior can realize social causation (as we saw in Section 5.1.2). If it seems these concessions eliminate the reality of social entities or are otherwise inconsistent, consider parallel cases from outside the social sciences. Organisms are composed of molecules and do not act independently of them; there are good reasons, nonetheless, for thinking biological theories are irreducible. Similar conclusions probably hold for ordinary macroscopic objects. So if social entities are not real because they supervene on and are exhausted by individuals, then organisms and chairs are not real either. Holists will be satisfied to grant that social entities are "only" as real as this.

A final individualist worry about holist ontology is that social entities

[36] See Kincaid (in press a, in press b) for a discussion of some of these issues.

are conventional and thus mind dependent in a way that real entities are not. Obviously this skeptical claim does not follow merely from the fact that the social sciences deal with mind-dependent entities like norms, conventions, and beliefs. Perhaps how we individuate the world into social groups may sometimes be partly "conventional" in that there may be multiple ways to divide up the social world. But that is true for galaxies, species, and some organisms as well, yet no one (except a philosopher) would claim that their existence depends upon being perceived.

Hence we have a strong prima facie case, based on the results of earlier sections, that social entities are a real part of the world.

5.4 The truth in individualism

We have now surveyed numerous individualist theses. We have seen that in most of its guises individualism is highly implausible. The conceptual arguments for reduction fail, and there is plenty of empirical evidence that good social theory can be irreducible. Attempts to find plausible non-reductive versions of individualism fare no better: individualist explanation case by case is drastically incomplete, there is no compelling evidence that individualist theories explain better than holist accounts, no general demand for individualist mechanisms is plausible, and even individualist heuristics are suspect in many instances.

What, then, is the truth in individualism? Three plausible theses have emerged from our discussion:

(1) the ontological thesis that society is composed of and does not act independently of individuals;
(2) the explanatory thesis that *some* reference to individuals is necessary for *full* explanation; and
(3) the evidential thesis that individualist mechanisms are sometimes essential to confirmation and when not essential they nonetheless contribute to confirmation.

While these theses are plausible, they also leave us with an anemic form of individualism. As we saw earlier, the ontological thesis is one no holist need deny, for it is compatible with the failure of reduction, the explanatory impotence of individualist theories, and so on. The evidential thesis is not much more controversial. Since presumably it is always better to have more rather than less evidence, no holist should deny that evidence about individual mechanisms adds *something*. Yet that leaves no special role for information about individuals.

Requiring some reference to individuals to explain fully is slightly more controversial.[37] Nonetheless, holists can easily accept that thesis as interpreted here. If a full explanation is a complete account of the causes, then reference to individuals will make any purely holist account more explanatory. Still, that admission does not derive holist accounts of explanatory worth, nor does it impose any significant constraints on holist theories. After all, the theory of evolution would be a more complete account in this sense if we could specify the underlying quantum-mechanical details. But we can hardly fault the theory of evolution for being seriously incomplete on such grounds.

Paige's work nicely illustrates these points. His social classes are of course made up of concrete individuals, and the actions of classes flow from the actions of individuals. His broad macrosociological claims are confirmed in their own right, but he also shows that they make good sense given what we know about the underlying individual behavior that brings them about. So he provides mechanisms to help confirm and explain. Yet Paige's work is firmly in the classical holist tradition: it explains by appeal to apparently irreducible social entities, social processes, and social laws; it explains the behavior of individuals by their social position; it appeals to evidence about macrosociological processes; and so on. This shows that only when individualism has little bite does it say anything plausible.

Thus we are left with a picture of whole and part or macro and micro in the social realm that is, I would suggest, typical of how science proceeds in general. Investigation, confirmation, and explanation proceed successfully at the macrolevel. Accounts at the level of the whole pick out patterns that may bear no neat relationship to the way lower-level accounts divide up the world, so reducibility is the exception rather than the rule. Wholes are, of course, made up of their parts and do not act independently of them. Evidence about underlying details can indeed be useful, but as a complement to rather than a substitute for more holistic information. Individualism, insofar as it says anything controversial, denies this general understanding, urging the social sciences to adopt an implausible and outmoded model of science.

---

[37] Putnam (1981) denies that these partial explanations explain at all, because they provide irrelevant detail and hide essential features. We can agree with Putnam that they do not capture regularities and patterns found at higher levels. But that does not mean they answer no questions. In particular, they might still partially answer the question "How did this event come about?" If they provide irrelevant detail, then they are inadequate as individualist explanations in the first place. But that does not show that we can never explain at all by describing individual-level detail. I have developed these arguments in considerably more detail in Kincaid (1988).

6

A science of interpretation?

The previous five chapters made a detailed case for a science of society. Nonetheless, important doubts remain unanswered – in particular, doubts arising from the fact that human behavior is meaningful. This chapter takes up these doubts; Chapter 7 examines the fact that the most developed social science, economics, seemingly has had very little empirical success.

Physical objects do not have a point of view on the world, do not attribute significance to the world around them, and more generally do not "mean." The physical sciences banished such subjective qualities long ago. Yet these features seem part and parcel of social explanation. This apparently fundamental difference between the social and the natural sciences has led many to argue that no science of society – at least in anything like its current form – is possible. Interestingly enough, philosophers from two very different traditions advance this skeptical thesis. Drawing on Dilthey, Heidegger, and Husserl, contemporary philosophers in the hermeneutical tradition such as Charles Taylor argue that (1) human behavior must be understood as meaningful; (2) understanding meaning requires interpretation, a process not amenable to naturalistic methods; and thus (3) no naturalistic science of society is possible. Though details no doubt differ, philosophers from the positivist tradition espouse similar arguments. Both Quine and Rosenberg think there can be no science of meaning and thus no social science as we now know it. Thus the two camps differ only on their reaction to the skeptical conclusion: Quine and Rosenberg want to eliminate social science, Taylor and the hermeneutic tradition want to eliminate naturalistic standards.

So the meaningful nature of human behavior seems a serious threat to naturalism. Recall that naturalism asserts two basic propositions: (1) the social sciences can be good science by the standards of the natural sciences and (2) the social sciences can be good science only by meeting the standards of the natural sciences. Arguments from meaning directly attack the first thesis, especially since I have defended social science as it is now practiced, not some drastically revised discipline like that advocated by Quine or Rosenberg. Hermeneuts like Taylor will reject the second thesis as well, for they believe there are ways of knowing unique to social investigation – standards for confirmation and explanation that are not instances of the basic scientific virtues typical of the natural sciences.

The argument of this chapter follows a familiar pattern. I shall first argue that meaning poses no obstacle in principle to well-confirmed explanations in the social sciences; I do so by defusing the relatively *a priori* arguments of Quine, Rosenberg, Taylor, et al. I then go on to look at some actual social research involving "meaningful behavior," namely the work of Brown and Harris on depression. That work, I shall argue, meets basic requirements for good science – more specifically, it provides causal explanations that have undergone fair tests, independent tests, and cross tests. Section 6.1 sorts out issues and then attacks a presupposition essential to the arguments from both camps – namely, that current social sciences as practiced now must be about meaningful behavior. Sections 6.2–6.4 explain and reject the arguments of Quine, Rosenberg, Taylor, and others, and then examine actual interpretive social science. Finally, Section 6.5 looks at difficulties associated with two specific kinds of interpretive social science – work that appeals to norms and work involving symbolic meaning.

6.1 Issues and presuppositions

To evaluate arguments from meaning, we need to explain more clearly the claim that human behavior is meaningful. "Meaning" is a notoriously difficult term to clarify (a fact which in large part explains Quine's skepticism). However, a useful first start is to distinguish the different concepts that social scientists invoke. At least the following notions are sometimes involved in the claim that human behavior is meaningful:

> *Perceptual meaning:* how a subject perceives the world, including the actions of others and him- or herself.
> *Doxastic meaning:* what a subject believes.
> *Intentional meaning:* what a subject intends, desires, etc., to bring about.
> *Linguistic meaning:* how the verbal behavior of the subject is to be translated.
> *Symbolic meaning:* what the behavior of the subject – verbal or non-verbal – symbolizes.

A science of interpretation? 193

Normative meaning: what norms the behavior of the individual reflects or embodies.

The first four categories are commonsensical enough, at least on the surface. Symbolic and normative meaning are less so. Symbolic meaning refers, for example, to the meaning anthropologists sometimes attribute to rituals – as when Turner claims that the Ndembu milk tree stands for matriarchy or Geertz claims that the Balinese cockfight is a comment on Balinese class relations. Normative meaning is invoked, for example, when an action is labelled as fulfilling a role, violating a rule, or reflecting a culture and its norms. No doubt these divisions are imprecise and their own meanings still obscure. But they will largely do for our purposes, and we can seek further precision when necessary.

So critics of naturalism claim that meaning in some or all of these senses precludes a science of society. Any argument along these lines rests on a presupposition that we need to make explicit, namely, that the social sciences *must* use meanings to explain. We can take this presupposition in several different ways: (a) *some* social research may be about meaning or *all* may be and (b) *full* explanations may require reference to meaning or even *partial* explanations may do so. The antinaturalist argument from meaning is most interesting if it asserts that *all* social science must refer to meanings to explain even partially. Weaker positions allow some social research to explain fully without meanings. The wider that domain, the less meaning threatens the naturalist program.

This presupposition of the arguments from meaning is troublesome. It is hard to show what any science *must* be like, come what may. Of course, one can always make the argument that without appeal to meaning one is not doing social science (Greenwood 1991, p. 27). But that trivializes the matter, for the claim then becomes that without reference to meaning, we cannot explain in terms of meaning – not much of an argument. (If this point is not clear, imagine a similar argument against modern chemistry by the friends of phlogiston.) Ultimately questions about the "proper" domain of a science are empirical questions within science itself. Really only a fully developed, well-confirmed social science that referred essentially to meanings would establish that social science must be about meaning. Even that claim would be fallible and in principle revisable.

However, *current* social science is not uniformly about meanings. Nor does it uniformly explain even *partly* by appeals to meanings. The work of Hannan and Freeman discussed in Chapter 4 is an obvious example. They explained the kinds of organization and their distribution. Since their explanations were at the level of social entities, not individuals, no essential assumptions about individual beliefs, perceptions, or intentions are involved. Such assumptions *could* be added in the process of providing individualist mechanisms for Hannan and Freeman's large-scale processes. But as I argued in the last chapter, large-scale explanations can succeed on their own – they can cite real causal processes – while remaining agnostic about the underlying detail. So Hannan and Freeman's functionalist accounts explain without any particular assumptions about belief, desire, and perception. In fact, they are compatible with an account of individual behavior that eschews such notions altogether. Nor, so far as I can see, do their explanations make any essential appeal to norms or symbolic meaning, though again we could imagine supplementing them with such factors.

So arguments from meaning have a restricted force, even if they are successful.[1] *They belie an individualist bias* – they assume that all social explanation must be about individual behavior. Yet when social science is about the large scale, it can abstract from individual detail and thus from explanations in terms of meaning. Of course, holists will count this feature as an argument in their favor, at least to the extent that meaning does raise obstacles to naturalistic investigation. Once again we see that naturalism and holism motivate and mutually support each other in the social sciences.

In what follows I shall give critics of naturalism their presupposition about meaning. I do so (1) because critics might still produce arguments that even work like Hannan and Freeman's presupposes meanings in some sense, (2) because my case is stronger if I can defend naturalism at both the social and individual levels, and (3) for the sake of the argument. Nonetheless, it is worth keeping in mind that arguments from meaning are a drastic threat to naturalism only to the extent that they make the strong and implausible presupposition that most social explanations invoke meanings.

6.2 The right-wing attack

In this section I survey and reject arguments from those in the positivist tradition, primarily arguments from Quine and Rosenberg. I label these criticisms "right wing" because they are about conserving standards against an onslaught of sloppiness and imprecision. Both Quine and Rosenberg apparently think a social science is possible, but not in anything like its current form. And both think the culprit lies in explanation via meanings. In Quine's case, the primary objections are to linguistic

---

[1] In his more recent writings Rosenberg (1988) acknowledges this more limited scope for his criticisms.

meaning; Rosenberg's worries are about attributing a belief and intentions. I begin with Quine's objections.

Quine has very little to say directly about the social sciences. Yet he says much by implication, for he believes that accounts invoking meanings and mental states more generally are "scarcely deserving of the name explanation" and "extravagantly perverse" (1975, pp. 87, 85). When it comes to meanings, there is not "an objective matter to be right or wrong about" (1960, p. 73). Commentators sympathetic to Quine have drawn the apparently obvious conclusions about the social sciences: Quine's arguments show that much social science is "empirically indefensible" (Soles 1984, p. 482); that attributions of meaning are "prescriptive" or normative, not descriptive (Feleppa 1988); and that attributions of meaning by social scientists are as much a matter of "imposition" as they are of "insight" or comprehension (Roth 1986a).

Quine's arguments for these conclusions are compressed and subtle (or should we say "sketchy and vague"?). I shall not provide detailed evidence about Quine's *true argument* – a notion of dubious sense if Quine's views are right – but shall rather develop several "Quinean" lines of reasoning.[2] These rationales fall into two basic categories: those turning on the kind of precise language a science requires and those turning on facts about translation. I begin with the former.

Quine inherits from the positivist tradition the notion that any adequate science must meet certain formal requirements. A fully developed science is expressible in a canonical language where the primitive terms, definitions, and logical relations between propositions are laid bare. However, such a formalized system seems unavailable for a discipline that invokes meanings. Meanings have properties that wreak havoc with the logical structure that for Quine epitomizes the ideal science. Descriptions of meanings are "opaque" – they do not allow us to substitute into a statement terms that refer to identical entities. For example, suppose a biologist claims that "the lactose gene is activated when lactose stimulates binding of a protein to the corresponding repressor gene." If we later identify the lactose gene with sequence L of DNA base pairs, we can then infer as part of biological theory that L is activated when lactose binds to the repressor. In short, we can substitute identicals and preserve truth and logical relations. However, we cannot do the same when it comes to meanings. If an anthropologist knows that a particular tribe believes lightning to be divine, he or she cannot infer that the tribe believes atmospheric electrical discharges are divine – even though the two phenomena are one and the

---

[2] In what follows I have been helped by Kirk (1986), P. Roth (1986b, 1987), and Feleppa (1988).

same. So their belief about lightning under one description does not transfer to lightning under another description. Thus any theory describing meanings seems unamenable to precise formalization with a clearly identified logical structure.

Such "opacity" does not prove that any social theory invoking meanings is fatally flawed. One way to show this is to ask whether the natural sciences meet Quine's stricture. Ethology and the study of animal behavior certainly seem to be exceptions, for they explain in terms of cognitive states that also exhibit opacity. Furthermore, Quine's ideal of a formalized theory is probably unrealizable in physics as well. Opaque contexts are an instance of a more general phenomenon philosophers call "intensionality." Intensionality occurs whenever a set of statements is not "truth-functional" – when the truth of compound statements is not determined by the truth value of their components. "Lightning is divine" and "weather-induced electrical discharges are divine" have the same truth value, yet "the Azande believe lightning is divine" and "the Azande believe electrical discharges are divine" have different truth values. It is this failure to be "truth-functional" that makes rigorous formalization difficult. However, there are good arguments that any time we make references to probabilities, to the future, or to what might be the case, the problem of intensionality arises.[3] Physics, however, is apparently rife with such references.

Even if opaque contexts were exclusive to the social sciences, we would still have to ask what difference that would make. Social scientists also know about opacity and have ways to handle it. It is well known, for example, that how a question is phrased can influence how it is answered, even when the different phrases describe the same thing. This means that we must be careful in attributing beliefs to individuals that go beyond what they explicitly say and must factor in how people's background beliefs influence the way they interpret questions. Good social science takes such steps.[4] However, these problems are just an instance of the more general fact that context is often important – a complication that is equally present in evolutionary biology, as we saw in Chapter 3. Thus, opacity is a difficulty, but not one that prevents good science or one that separates the social from the natural sciences.

A much deeper rationale for Quine's antipathy to meanings is his doctrine that translation is indeterminate. Imagine an anthropologist trying to learn and translate a language previously unencountered by outsiders – what Quine calls "radical translation." The anthropologist builds up an

---

[3] For a summary of such arguments see Hookaway (1988, ch. 6).
[4] Marsh (1982) discusses such problems in survey research and questionnaire design and the ways they can be handled.

A science of interpretation?

understanding of the language by observing what the natives say in what contexts, starting with simple terms for common objects and moving on to eventually build a dictionary for the whole language. Quine claims that there will always be multiple ways of constructing these dictionaries such that they are (1) equally consistent with the evidence and (2) in no way compatible or simply trivial variants of each other. This means, according to Quine, that there is no fact of the matter about whether two expressions mean the same thing. But if there is no fact of the matter about meanings, then there can be no science that invokes them.

As with medieval arguments for the existence of God, there are as many versions of Quine's reasoning as there are commentators. However, some points about this general strategy seem clear enough. Quine is claiming more than just that the evidence does not uniquely pick out a translation, for he believes that all theories are underdetermined by the evidence. So Quine's claim is not just epistemological – it is also an ontological claim. The ontological component comes from his physicalism. For Quine, the physical facts are all the facts. So to say there is no fact of the matter about meaning is to say that alleged differences in meaning do not show up as physical differences.

I shall thus put Quine's general argument this way:

(1) The physical facts do not determine the semantic facts.
(2) The physical facts are all the facts.
(3) Therefore, there is no fact of the matter about meanings.
(4) Therefore, there is no right answer about questions of translation.
(5) Therefore, there can be no adequate science that invokes meanings.

The first premise is an ontological claim – it says that once all the physical facts are fixed, the semantic facts can still vary. Or, in terms used in the last chapter, the semantic facts do not supervene on the physical facts. Quine gets premise (1) from his story about radical translation, which he believes establishes that the totality of physical facts available to the translator is not enough to ensure one and only one translation. (2) is a component of Quine's physicalism and is what we called in the last chapter an "exhaustion" principle. (3) is an inference from (1) and (2), relying on what Quine means by "a fact of the matter." (4) concludes that since there is no fact of the matter, there can be no one "right" translation or attribution of meaning. (5) is then a further inference that draws morals about the social sciences from the indeterminacy of translation.

My main objections to Quine's argument fall into three broad categories:

Lack of evidence for the anti-determination claim. What is the argument for (1)? If Quine is to be consistent with his own epistemological views, then (1) must be a defeasible empirical claim, not a conceptual truth. Though Quine often acts as if his claims have the force of a necessity, he ultimately grants that they are empirical (1990, pp. 198–200). So (1) must be an empirical generalization. What then are the facts that base that generalization? Quine provides some minimal facts about the translation process and some few examples illustrating indeterminacy. This is hardly overwhelming evidence, especially since critics have raised doubts that Quine's examples really show indeterminacy (Kirk 1986). Moreover, Quine's account of translation gives a very impoverished description of the "physical facts." If we want to know whether the physical facts determine meaning, we have to consider *all* the physical facts. For example, anthropologists often use what they know about economic, social, and ecological circumstances to choose between translations. Quine would need to show that the physical facts underlying those processes left meanings indeterminate, something he does not do. He sticks to a very narrow subset of the physical facts – medium-sized physical objects described in common sense terms. (Quine claims that he is working with "stimulations at the sensory surfaces" but drops that unusable and vague notion when it comes to talking about real translational practice.) However, arguments for (1) from that restricted notion of the physical are not enough to show that *richer* sets of physical facts do not determine translation.

Doubts that "no fact" in Quine's sense entails no best translation. Quine's notion of "no fact of the matter" rests on his physicalism and is an ontological claim. Yet the indeterminacy claim is epistemological – it is that there is no unique, best-confirmed translation. As I have stylized the argument, (3) straddles the fence on these two readings: it can be read either as a claim about ontological determination or as a claim about evidence. But these two readings are importantly different. Since Quine's conclusion is epistemological, (3) must be read as about evidence. However, (1) and (2) do not tell us that there is no best translation, for they are ontological claims, not epistemological. Even if the physical did not determine the semantic, how do we know that there is not sufficient evidence to pick a unique translation? After all, the physical probably does not determine the mathematical either, but that surely does not tell us what is and is not decidable in mathematics. Though at times Quine claims to have shown that translations are not even fixed by all possible evidence, he has done no such thing unless the only possible evidence is physical evidence, and then physical evidence in his restricted sense. So establishing that there is no fact of the matter in Quine's ontological sense does not establish that there is no best translation.

A science of interpretation? 199

Quine's best move here is to bring in as a further factor the idea that theories are always underdetermined by their evidence. Then we can take (3) simply as an ontological claim, with the conclusion following from it plus the further assumption that all theories are undetermined by their evidence. Quine's argument would then say:

(1) The physical facts do not determine the semantic facts.
(2) The physical facts are all the facts.
(3) Therefore, there is no fact of the matter about meanings.
(4) All theories are underdetermined by their evidence.
(5) Therefore, any translation is underdetermined by the evidence, that is, the evidence is always compatible with more than one translation.
(6) Therefore, there is no right answer about questions of translation.
(7) Therefore, there can be no adequate science that invokes meanings.

(3) is now understood as an ontological claim, and there is no attempt to show indeterminacy simply from the fact that the physical facts do not determine the semantic facts. Epistemological morals come in only after we assume that theories, including translations, are underdetermined by their evidence. Thus in linguistics, unlike physics, there are multiple accounts compatible with the evidence *and* no fact to be right about. This reading does avoid the equivocation over "fact of the matter" described above. However, it does so at a price. The claim that theories are underdetermined by all possible evidence is a strong one indeed and smacks of conceptual proofs about what science can and cannot do. Moreover, persuasive arguments have recently been raised that Quine's notion of underdetermination looks plausible only because of his narrow and implausible notion of evidence (Laudan and Leplin 1991). So if the appeal to underdetermination is essential, then Quine's indeterminacy inherits its uncertainty.

Doubts that indeterminacy rules out objective social science. For our purposes, the key question is whether indeterminacy warrants skeptical conclusions about the social sciences. Imagine that it is indeed true that the best social science makes assumptions about meanings, contra the arguments of 6.1. Imagine that we have a well-established case that those meanings can be translated in two, incompatible ways, contra the doubts raised above. That would still not show that our social explanation was not objective for two reasons: (1) The same social account might hold regardless of which translation of meaning was used. To take an example from Quine, suppose the evidence gives us no way to decide between translating the native expression *gavagai* as "rabbit" and "associated rabbit stages." It might still well be that any generalization we make about social practices – say about rituals or hunting behavior – holds up however we translate that term. (2) The indeterminacy thesis is ambiguous, for it can be given either global or local readings. Construed locally, it claims there is at least one sentence that rival translations will assign different, incompatible meanings. Construed globally, rival translations disagree on all sentences. Quine no doubt intends his thesis to fall somewhere between these two extremes, that is, as something less than a global claim. So indeterminacy of translation is compatible with two rival translation manuals agreeing on a range of translations. But then there is room for objective social science invoking meanings and the like – namely, wherever rival translations agree. In short, indeterminacy need not infect all explanations via meanings and even where there is indeterminacy, good social science may still be possible.

Quine's move from indeterminacy to subjectivity is also implausible because it threatens to rule out much more than social science. For example, consider the biological category of "species" and Quine's notion of "behavioral dispositions." Neither, I would argue, is determined by the physical facts. Both are summary concepts that collate lower-level details; as a result, we can employ them in different ways, and nothing in the physical world nor in the evidence forces us to choose one application over another. There is an enormous debate among biologists about how to define the species concept. Some favor the "biological' concept wherein ability to cross-breed is definitive. Others delineate species by ecological roles. Paleontologists favor morphological accounts. As Kitcher (1984) has argued, there may be no "right" answer among these concepts; rather, they represent different ways of grouping organisms for different purposes. Should we therefore conclude that there can be no real science involving species? Quine's argument would seem to suggest so.

Similar problems arise for Quine's favored behaviorist concept of "stimulus meaning" – our dispositions to verbal behavior – which is neither a concept from physics nor uniquely determined by the physical facts. We can imagine categorizing behavior dispositions in two different ways that produce the same predictions because we have made compensating changes – providing fine individuations of behavior here, grouping stimuli differently there, and so on. Thus either indeterminacy alone is not enough to draw Quine's skeptical conclusions or even his own version of objective "semantic" concepts must go.

Thus indeterminacy by itself does not warrant skepticism about theories that invoke meanings. Deciding that two expressions should be equated involves individuating or grouping behavior. Nothing in the physical world may ensure that there is just one way to make such groupings. But that is likewise the case for categories in many higher-level sciences. However, that slack in no way implies that those categories are inappropriate for scientific explanation or inevitably subjective. Quine's arguments do not preclude a similar conclusion about meanings.

I want to turn next to Rosenberg's somewhat different attack. Rosenberg's worry is about "belief-desire psychology" – about explaining human behavior by appeal to mental states like preferences, beliefs, and so on. Rosenberg claims that these explanations rely on one basic principle, labelled [L]. Roughly the principle is this commonsensical one: if someone wants A and believes that doing B will bring about A, then that person will do B. Following Churchland (1979), Rosenberg believes this principle makes "folk-psychological" explanations unrefinable and scientifically inadequate.

Rosenberg, so far as I can see, has two main arguments for this conclusion:

(1) The argument from circularity (1988): attributing beliefs requires determining what people desire and how strongly, yet doing this presupposes we know what they believe. In short, every time we try to test folk psychology we must presuppose its truth. This shows folk psychology to be unrefinable, for no test will lead us to give up its basic principle.

(2) The argument from many models: The holistic nature of belief-desire attribution means that the same behavior can always be explained in multiple ways. We can always attribute different desires behind the same behavior by making corresponding changes in what people believe. In Rosenberg's words, "too many such systems are easily constructible, and they all have the same prima facie explanatory power" (1980, p. 81).

Two general problems confront both arguments. The first is simply that Rosenberg's conclusion is too strong; he claims to show us what any potential science employing belief-desire concepts must be like. I think we have good inductive evidence to suspect any such argument, for nearly all past attempts to legislate what investigation can and cannot do have failed. We also have good Quinean grounds sketched in Chapter 2 for doubting that broad conceptual considerations can tell us how science must work. The second problem is that Rosenberg's arguments threaten to go too far. Parts of current ethology, evolutionary biology, behavioral ecology, primatology, and so on are also doomed if Rosenberg's arguments succeed. They, too, explain in part via belief-desire or intensional descriptions more generally.

Of course Rosenberg could extend his skeptical attack to these domains as well. Yet biting the bullet in this way should increase the first worry mentioned above, for now we are legislating not just what the social sciences can or cannot successfully do but also making that judgment for important parts of biology.

Of course, these considerations are not decisive. Rosenberg can grant my Quinean philosophy of science but deny that his arguments are *a priori* in any objectionable sense. Rather, they are empirical explanations for why the social sciences have failed. However, Rosenberg's explanations are uncompelling. Consider first his argument from circularity. He claims that belief-desire generalizations, unlike the gas laws, are not independently refinable. We have no "belief thermometers," only inferences from other beliefs. However, this example leaves out a crucial fact, namely, that our understanding of how thermometers work, of what they measure, of when they accurately measure, and so on depends on statistical mechanics. But the gas laws are just instantiations of statistical mechanics itself. So at this level of description, the problem Rosenberg cites is not unique to belief-desire psychology: to measure the variables of statistical mechanics, we need statistical mechanics itself. However, statistical mechanics is refinable, because it is built from many different and independent hypotheses, assumptions, and so on. As we saw in Chapter 2, such complexity allows a theory to be used in testing itself without any troublesome circularity.

So the crucial question is whether belief-desire psychology exhibits sufficient internal complexity to ground real tests. I think the answer is affirmative. "Belief" and "desire" are catchall terms for a host of different categories employed in real psychological explanation. Short- and long-term memory, reasoning heuristics, classification prototypes, motivation states, and the like are just a few of the many different states that fall under the "belief" and "desire" headings. Aside from these different *types* of states, real psychological explanations invoke many different *token* mental states as well. For example, explaining voting behavior involves not just "attitudes" but attitudes about specific objects, e.g. towards specific elected officials, political parties, race, and on and on. So belief-desire explanations involve numerous separate and independent claims, making real tests possible.[5]

Rosenberg's best response to these arguments is that all such testing nonetheless assumes the general principle that if someone desires A and believes doing B will bring about A, they will do B. However, nothing

---

[5] For some real examples of how this works, see Henderson (1991, 1993). My discussion in this section has been much helped by Henderson's critique of Rosenberg.

significant follows. All testing in science presupposes that the main categories of the theory are the relevant factors for explanation. For example, when a field test of natural selection does not show survival of the fittest, biologists go back and look for missed traits and environmental conditions, ask if they have improperly identified and individuated traits, and so on; in short, they presuppose the fittest survive.[6] Yet evolutionary biology is tested and refined, for the principle "the fittest survive" does not capture the real content of evolutionary explanations and testing. As I emphasized earlier, real explanation and testing are to be found in the details. It is there that *ceteris paribus* clauses are filled in and concrete causes identified. So to focus on "the fittest survive" is to confuse evolutionary explanations with a slogan, when it is the many quite specific claims about particular environments, traits, life histories, and so on that do the real scientific work. Focusing on principle [L] is doing much the same. Of course at some point theories have to be willing to pay up – to grant that their basic categories do not provide successful or superior explanations or to grant that those categories must be revised. But what makes that reckoning possible is the specific claims, tests, and explanations that embody those categories. And presupposing that one's basic categories apply does not make such testing impossible.

Rosenberg also argues that belief-desire psychology fails because it allows too many explanations. By making compensating changes in the total system of beliefs and desires, we can produce multiple explanations of the same behavior. This problem does not really depend on the belief-desire circle – even if we had a "desire meter" that directly measured desires, we could still attribute multiple different *beliefs* all compatible with a given behavior. So Rosenberg's argument point looks much like the claim that since testing is holistic, theories are underdetermined by the evidence. Or, to put the basic idea in more traditional philosophical terms, if we have no direct access to reality but must infer its nature from indicators of some sort, we have no guarantee those inferences are correct. In short, Rosenberg's worries are a variant of traditional skepticism.

There are at least two fruitful routes to take against such skeptical arguments: to deny their premises or to deny their significance if successful. It is not at all obvious that many different belief-desire models are always possible. A compelling argument for this presupposition would look at real psychological explanation at work and then actually construct a competing model that (1) was not a trivial variant and (2) had the same empirical warrant. Rosenberg does not construct, so far as I know, such real

---

[6] Such moves in the face of falsifying evidence are nicely described by Endler (1986).

counterexamples. To that extent his argument is uncompelling – it no more shows belief-desire psychology must be inadequate than the abstract possibility that modern physics is underdetermined by the evidence shows that it is scientifically inadequate.

Even if we had some real cases of equally compelling but incompatible models, dire consequences for naturalism need not follow. If multiple models were widespread and persistent over time, then skeptical conclusions might be warranted. However, occasional instances where competing models fit the data would be no cause for alarm, for this is the stuff of ordinary science and a situation that simply calls for further search for data. Thus only a detailed look at actual psychological investigation could show whether multiple models are actually a problem.

Furthermore, multiple models consistent with the evidence is not necessarily a fatal flaw, as we saw when discussing Quine's indeterminacy doctrine. Important causal explanations of individual or group behavior may hold on all reasonable models of individual belief and desire. Political scientists, for example, have amassed a large body of data showing that socioeconomic variables influence voting behavior. Exactly how that process is represented in attitudes and other mental states is a matter of debate. Yet multiple explanations of this psychological mechanism need not undermine the well-confirmed causal influence of social factors, for those models can be consistent with that fact. A similar situation occurs in work on kinship in anthropology: we have fairly good evidence about large-scale patterns and multiple plausible models of the underlying belief structure involved.[7] Again multiple models of underlying psychology do not prevent good social explanation.

Thus neither Rosenberg nor Quine has shown inevitable obstacles to a science of society.[8] Nonetheless, we can appreciate their worries without adopting their radical eliminativist attitude. Sections 6.4 and 6.5 will discuss some of the real practical difficulties that meaning brings to the social sciences, and many of those problems will be ones like those identified by Quine and Rosenberg. Before we take up actual interpretivist social sci-

---

[7] For a review of some relevant data, see Levinson and Malone (1984).
[8] I should note that Rosenberg's arguments are inspired by Davidson's claim that the holism of interpretation commits us to a strong principle of charity and that in turn ensures that there is a normative element in interpretive science not found in natural science. I am unconvinced by Davidson's claims, for any reasonable principle of charity is a contingent empirical truth about what best explains, thus grounding no distinction between the natural and social sciences. For a decisive and exhaustive critique of Davidson and other arguments from the principle of charity, see Henderson (1993). See also Fodor and LePore (1992, chs. 3 and 5).

A science of interpretation?

ence, however, we have next to consider a second set of arguments from a rather different direction.

6.3 Skeptical hermeneuts

Critics like Quine and Rosenberg are skeptical of naturalism, not in principle but in practice – they believe that the social sciences will never meet standards of scientific adequacy so long as they deal with meanings. Their skepticism is shared by a quite different tradition which draws quite different conclusions. I have in mind the hermeneutical or interpretivist account of the social sciences whose best-known early representative was Dilthey (1989) and whose modern advocates include Charles Taylor (1971, 1980), Anthony Giddens (1976), Jurgen Habermas (1985), and others. Like Quine and Rosenberg, these philosophers and social scientists also deny that the social sciences can meet the basic standards of scientific adequacy. They do not, however, conclude that the social sciences produce no reliable knowledge. Instead, they infer that the social sciences need not meet naturalistic standards, for the social sciences have their own routes to knowledge which in no way embody the scientific virtues outlined in Chapter 2. If Quine and Rosenberg are conservatives out to preserve standards, the hermeneutical tradition glories in a left-wing pluralism.

In what follows I answer this "left-wing" attack, focusing on the work of Charles Taylor. Like Quine and Rosenberg, Taylor tries to find some universal and essential fact about the social sciences that shows them unable to meet naturalistic standards. His arguments start, of course, from the fact that human behavior is meaningful. However, it is the intervening premises – the claims about explaining meaningful behavior – that are crucial to drawing antinaturalist conclusions. Below I look at arguments based on the alleged facts that (1) understanding meaning relies on prior understanding, (2) meanings can be understood in multiple ways because interpreters may differ in their background assumptions, (3) interpretation requires reference to the subject's own categories, and (4) interpretation is contextual. While these arguments come from Taylor, they are widely advanced by others. None of them shows any inevitable divide between the natural and social sciences.

Drawing on an insight gleaned from Heidegger and the phenomenological tradition, Taylor argues that interpretation – the process of understanding meaning – relies on a prior understanding. Interpreting an individual's behavior, speech, beliefs, perceptions, and so on requires a prior set of categories, beliefs, and intuitions about what is appropriate behavior in a given circumstance and other such assumptions. According to Taylor, this means that the social sciences cannot have "brute" or neutral data as do the natural sciences. All interpretation involves a hermeneutic circle that begins with a prior set of meanings and then uses them to determine the meaning of other people's actions. So investigating and explaining meaningful behavior must be a very different process from that involved in confirming hypotheses by experimental test against objective data.

As should be obvious from the discussion in Chapter 2, this argument rests on an untenable assumption about natural science. Quine's doctrine that testing is holistic and Kuhn's emphasis on the role of background assumptions in natural science show decisively that natural science also depends on a prior understanding and background assumptions. How the data are described, what data are relevant, what the data imply, what questions need to be asked, and what hypotheses are worth pursuing all depend on prior beliefs, categories, and assumptions. So there is nothing special about meaning in this regard – a point that Taylor and other defenders of this argument should have recognized, since it is made by Ur-hermeneutists like Gadamer (1977).

Obviously, interpretivists need something more to make their argument work. Perhaps we might strengthen their argument by adding a further assumption: that interpretation is relative not only to background assumptions but that those assumptions vary. Natural science, the argument might go, does indeed rely on background assumptions. Those assumptions, however, can themselves be adjudicated through further evidence. The interpretative sciences, on the other hand, are much more open ended. Multiple background assumptions are possible and probable – all equally compatible with the evidence. It is human interests that determine background perspectives, not the data. Herein lies the difference with the natural sciences.

Taylor is of course right that social scientists may approach the same meaningful behavior with different background assumptions and thus produce different interpretations. Yet this fact does not inherently divide the social from the natural sciences. Obviously natural scientists sometimes approach the same phenomena with different theoretical presumptions and different resulting interpretations, as Kuhn has so forcefully pointed out. However, in the natural sciences these disputes are ideally resolved by conducting what we have called fair tests. So Taylor needs to give us an argument showing that such fair tests are impossible in the social sciences, something he does not do.[9]

---

[9] Another version of Taylor's argument would make the key factor the *implicit* nature of background knowledge in the social sciences. As Rouse (1987) nicely points out, this argument fails because implicit background knowledge is part and parcel of natural science as well.

A science of interpretation?

Antinaturalist conclusions are sometimes drawn on the ground that interpreting meaningful behavior requires reference to the subject's own concepts, perceptions, attributions of meaning, and so on. This claim has a long pedigree, yet exactly what it asserts and how it supports the antinaturalist position is unclear. Note that we first need to ask whether reference to the agent's understanding is essential as *evidence* or essential for *explanation*. Next we need to ask exactly why reference to the subject's understanding makes the social sciences different. I can discern three rationales: (1) that grasping the agent's conceptions requires special methods that are not subject to the traits of good science discussed in Chapter 2, (2) that appeal to the subject's conceptions makes social science subjective, and (3) that appeal to the subject's categories makes the social sciences strongly contextual. These claims and arguments are surely distinct; evaluating this traditional argument requires keeping these differences in mind.

We can surely sometimes explain without invoking the subject's categories and perceptions. When we are interested in explaining large-scale processes and patterns, the individual's beliefs constitute an underlying mechanism. However, we have already seen that citing mechanisms is not essential to explain, and thus a similar conclusion will sometimes hold for the subject's meaning. Moreover, we have good empirical evidence that individuals' beliefs are sometimes irrelevant to their behavior – that their behavior is at odds with their beliefs (see Liska 1975). Studies on attitudes and behavior concerning discrimination, health practices, nationalism, and child rearing suggest that attitudes may have little causal influence on behavior in these cases. If so, then what the subject believes cannot be essential to explaining behavior. *A priori* constraints like those imposed by Taylor legislate without warrant how empirical investigation must go. Even if the empirical work cited above turned out to be wrong, surely it is no *a priori* truth that it must be so. Hence, we have good reason to think that the interpretivist's stricture is not a *necessary* truth and that in some circumstances it is not even a contingent one.

Let's turn now to arguments about evidence. We can grant the claim that the agent's self-conception is potentially relevant evidence. Given the holism of testing, this claim is true but trivial – since potentially any piece of information could be relevant evidence in the right situation. Stronger claims are unwarranted. As I mentioned above, agents can be confused about the reasons for their behavior. So we might well know on empirical grounds that the subject's beliefs are in fact *bad* evidence. Of course, the subject's perceptions may be important and probative in some cases. That conclusion, however, is far short of the interpretivist claim that appeal to beliefs is essential.

Even if the subject's meanings were essential evidence, it would be unclear that the desired conclusion – that there is a sharp divide between the social and natural sciences – follows. Ethology and evolutionary biology also must at times know the meaning their subjects attribute to the situation. Cheney and Seyfarth's *How Monkeys See the World* (1990) is a compelling illustration; it provides explanations of vervet behavior that appeal essentially to how they interpret the world, backed by a rich body of evidence. So we have no sharp line here between the social and natural sciences.

Traditionally, interpretivists have claimed that evidence about a subject's categories and beliefs is obtained by some special process of insight unknown in the natural sciences. Previous defenders of naturalism have denied that insight or imaginative reconstruction counts as evidence at all (Abel 1948). However, naturalists need not go so far, especially since we obviously use such information frequently. Insight and imaginative reconstruction can count as evidence, but to do so they must meet the basic requirements of good science. In short, insight can be evidence, but only because it embodies the basic scientific virtues described in Chapter 2. If my imaginative reconstruction is heavily dependent on the theory I am trying to support, then my reconstruction counts for little, since it is eschewing fair and independent tests. Insight must also be cross checked. Imagine that I use my own beliefs, experiences, and so on to produce an initially plausible hypothesis about the beliefs of another. That evidence must be followed by other tests. I want to know how my hypothesis fits with what else I know about the subject's behavior and the social environment. If I postulate a set of beliefs which do not fit subsequent behavior and social practices, the evidence from my imaginative reconstruction will be overridden by what else I know. So insight may provide evidence, but it is evidence that must be checked by cross testing.

If we look at the practice – rather than the rhetoric – of interpretivist social science, we find exactly this more limited role for insight. For example, Geertz is a well-known advocate of interpretative social science and of the claim that interpretation differs sharply from the causal analysis typical of natural science. Nonetheless, Geertz's ideology and his practice are at odds. In *Kinship in Bali* (1973c) Geertz describes in detail how the Balinese see the world and what their institutions mean. To do so, however, he must likewise make standard naturalist judgments about the causal structure of Balinese society. For example, in arguing for his interpretation Geertz makes assumptions about Balinese political structure, about the social effects of wealth, about causal processes producing competitive struggles over access to social roles, about the social forces promoting cohesion, about the effects of kinship groups on politics, and so on. Understanding what people mean and understanding what forces shape their

A science of interpretation? 209

world go hand in hand. So "grasping the point" (Winch 1985, p. 115) may involve evidence based on insight, but Geertz tests that initial evidence by the usual methods of good science – and he of course invokes standard naturalist causal explanations in the process.

Another route to the antinaturalist conclusion comes from arguing that including the subject's perspective makes interpretation unacceptably subjective. According to Taylor, natural science operates under a "requirement of absoluteness." That means in Taylor's words that "the task of science is to give an account of the world as it is independently of the meanings it might have for human subjects" (1980, p. 31). Taylor's requirement is less than clear, but we can see the basic idea: the content of science should reflect the way the world is, not the peculiarities of our experience. In short, it should be objective. The elimination of qualitative features like colors from physics during the scientific revolution, for example, is the kind of thing Taylor has in mind.

We can grant Taylor's requirement. Yet it is hard to see how it rules out good science in the study of meaningful behavior. In fact, his argument turns on such a simple equivocation that charity would call for a different interpretation if one could be found. The equivocation comes in sliding from "as it appears to the subject" to "is subjective." The two notions are obviously different, and there is no reason why we cannot have an objective account of the way things seem. The mere fact that a description is about how someone interprets the world does not entail that it is subjective in the sense of not factual. The "absoluteness requirement" is not violated simply because we are describing subjects.

At times Taylor suggests that it is the *contextual* nature of interpretation that separates it from the natural sciences. Natural science claims are universal (hence "absolute"), but interpretations are "in terms of the situations to which they are essentially related" (1980, p. 35). We saw similar worries about laws and generalizations in the social sciences in Chapter 3. And the same basic reply will suffice here: contextual factors play an important role in much good natural science and are no inherent obstacle to confirmation and explanation. Numerous claims and concepts in the natural sciences have an inherent reference to a "context," with fitness being a prime example. Confirmation and explanation in the natural sciences are likewise often a piecemeal affair where unique variables and conditions must be factored before hypotheses are well confirmed. Once again the interpretivist argument turns on a simplistic view of the natural sciences.

No doubt there are further ways to use meaning to argue for the inevitable difference between the natural sciences and the "human" sciences. I will end my survey of such arguments here, trusting that the above sample warrants a reasonable generalization: no argument based on the meaningful nature of human behavior is likely to show what the social sciences can or can never do. Our next task is thus to look at interpretive research in practice and to argue that good work is possible in practice, not just in principle. But before turning to that task, I want to conclude this section by doing two things: drawing some morals about the kind of practical obstacles facing any science of interpretation and forestalling potential confusions about what the arguments given so far have shown.

I have defended social theories that trade in meaning. Yet that does not entail that I am committed to every way those theories are *interpreted*. In particular, my defense does not commit us to a strong realist view about meaning and psychological attribution. Behind Quine's attack on meaning are ontological worries.[10] Equating two expressions because they share the same meaning seemed to him extravagant Platonism; in other words, it appeared to commit us to entities called "meanings" that are non-physical things. Doubts like Rosenberg's can also be motivated by worries about realism. When we ascribe beliefs and desires, should we take those descriptions realistically? Should we assume that there is one psychological state corresponding to each belief we ascribe? That those states are somehow language-like? Nothing in my defense of interpretive social science commits us to these realist construals. We can find expressions equivalent and think we have good reason for doing so, yet not commit ourselves to meanings as entities – either as abstract entities along the lines of propositions or as mental entities along the lines of "representations." To take an example from biology, geneticists made good use of the notion of "having the same gene" long before they had any idea what the genetic mechanism was – and explaining by reference to genes did not commit them to any elaborate theory of biochemical processes. So too with meaning in the social sciences: social scientists may provide well-confirmed explanations involving meanings without committing themselves either to meanings as abstract entities or to any very elaborate psychological theory involving representations or a language of thought, for example.[11]

My defense of interpretive social science also does not entail that all, most, or even much interpretive social science is good science. If the arguments from meaning fail, they nonetheless point to potential, albeit contingent, obstacles to good science in the social sciences. As we turn to actual practice in the next section, it will be important to keep such prob-

---

[10] See P. Roth (1987, ch. 6) for an exposition of this side of Quine and for a discussion of how attacks on realism might defuse some standard controversies in the social sciences.

[11] The point made here about meanings as entities I take from Kirk (1986) and that about representations from Horgan and Graham (1991).

lems in mind. From the arguments of the previous sections we can glean the following reminders:

(1) Belief contexts are opaque and thus attributing beliefs is by implication risky business.
(2) Subject categories are potentially relevant evidence and thus social scientists ignore them at their peril; on the other hand, uncritical acceptance of the subject's account has equal dangers.
(3) Beliefs, perceptions, and so on are often contextual; thus generalizing across contexts is likely to lead to errors.
(4) Attributions of beliefs are easily underdetermined by the evidence; this means that alternative interpretations must be considered and ruled out – in other words, fair tests are particularly important.
(5) In light of the above point, simply describing how a society "sees the world" may provide little explanation. If we take explanation as essentially about citing causes, then accounts of meaning provide explanations only if they are part of well-confirmed causal accounts. When there are multiple possible plausible interpretations, then we may have no ground for taking any one of them as identifying a real psychological state. In that situation the beliefs ascribed cannot be reasonably claimed to be the causes of behavior and thus – if explanation requires citing causes – we cannot explain.
(6) Attributing meanings is done holistically; the smaller the circle of beliefs and desires, the less chance there is for meaningful independent tests; correspondingly, interpretations are more likely to be compelling if they are integrated with a full-bodied account of social processes and structures.
(7) Attributions of meaning can be "thick" or "thin" – in other words, can involve highly detailed claims about what people believe, their cognitive processes, and so on, or can involve much more minimal, generic assumptions about meaning. Strong evidence for a thin description does not automatically carry over to thicker accounts; doubts about thicker accounts need not disparage thinner ones.

Each of these morals arises from the arguments considered in the last two sections. They are the real insights behind the arguments of Quine, Rosenberg, and Taylor – and, as we shall see, they are lessons that interpretive social science ignores only at its peril.

212   *Philosophical foundations of the social sciences*

6.4 Interpretive successes

My goal throughout this book has been not just to argue that good social science is possible in principle but to argue that some current social research actually meets basic standards of scientific adequacy. Though I deny meaning constitutes an inherent obstacle, it does constitute a real practical one, and social science is, on my view, often at its weakest when it is most interpretive. Moreover, since my concern is to defend a science of the large scale, I will not here argue that there is a significant body of well-confirmed interpretive social science. However, good interpretive social science does exist, and in this section I discuss one such piece of research that nicely illustrates how social scientists can handle doubts about meaning in practice.

In the *Social Origins of Depression* (1978) Brown and Harris take up an ambitious task – to develop and test a theory about the social causes of depression. Earlier work had suggested that depression is not evenly distributed across social groups. Based on a variety of clinical evidence I will not go into, Brown and Harris proposed the hypothesis that social class influences depression by two routes – by increasing the odds of certain life events that provoke depression and by lowering the odds that individuals have certain "protective" social relationships such as employment and a stable family relationship.

Brown and Harris test their hypotheses by in effect running a retrospective clinical trial. A retrospective trial tests a hypothesis by comparing those with a given condition to a control group without the condition. If the experimental group has characteristics unseen in the control group, then those characteristics are possible causes. If those characteristics occur at a rate higher than would be found by chance, then we have stronger evidence that they are causal factors, for we can rule out the hypothesis that our data result from random sampling differences. If the same characteristics clearly occur prior to the conditions of interest, then we have stronger evidence yet that the characteristics in question may be causal factors. And if steps are taken to ensure that the control and experimental groups are similar on all other factors – or if such differences are controlled statistically – then our confidence that we have found a causal factor also increases.

The Brown and Harris study takes all of these steps and more. They compare three groups: 114 psychiatric patients classified as depressive, 76 non-psychiatric patients diagnosed with depression, and 382 randomly selected controls without depression. Two "treatment" groups are used to be sure that factors influencing who seeks medical care do not confound the results. The number of life events and the extent of "protective" social ties are measured in each group, as is the frequency of these factors by

A science of interpretation? 213

social class. If Brown and Harris are right that social class causes life events and life events cause depression, then we should see greater probability of life events in the depressed groups, with a disproportionate class representation as well.

Brown and Harris's test results strongly support their hypotheses about the social causes of depression, and their study design helps rule out obvious competing explanations. As social research goes, there is little doubt that its design is exemplary. My main interest is whether this study successfully avoids the obstacles to good science raised by explanations invoking meanings. I think it does for three reasons.

First, Brown and Harris's hypotheses make minimal assumptions about their subjects' "true" underlying beliefs, cognitive states, and so on. They do not presuppose any specific psychological account of depression nor any specific psychological mechanism linking their independent social variables to their dependent variables. Their dependent variables are how individuals respond to questions – in short, verbal *behavior*. While they, of course, take those responses to indicate depression, they are not committed to a unitary entity "depression" or any elaborate belief-desire account of it. So Brown and Harris provide a "thin" account of meaning. Thus they can largely avoid concerns like those of Quine and Rosenberg about psychological validity, for they make only minimal assumptions about underlying psychological processes.

Second, Brown and Harris's evidence comes from interviews. Thus there are the concerns (a) that there are multiple ways to interpret those interviews and (b) that the interviews will be contaminated by the background expectations of the investigators who will reclassify any responses that violate their hypotheses. The first problem is Quine's worry about indeterminacy; the second, an instance of Rosenberg's worry about testability.

Like Quine, Brown and Harris also worry that there may be multiple ways to categorize the responses they get (as well as the independent variables). So they take steps like those pointed out in Section 6.2: (1) they use a thin account of depression, one that uses lowest-common-denominator criteria – criteria shared across different approaches to depression; (2) when a subject seems open to multiple interpretation – in short, when critics of their work might challenge classifying a particular subject as depressed – they exclude the case from their data; and (3) they examine whether their results hold up when different groupings of the responses are used. The first two procedures are ways of showing that, to use our earlier terminology, indeterminacy is not global – in short, that there are a range of cases on which competing interpretations would agree; the third procedure tries to show that even if there are multiple ways to assign meaning, the proposed causal relationship still holds.

Brown and Harris are likewise concerned about the testability issues bothering Rosenberg, and they take numerous steps to deal with them. They use outside interviewers, they set the standards for classification of responses prior to conducting the interview, they drop cases whose interpretation might beg the question against alternative hypotheses, they require numerous different responses on different topics before they classify a subject as depressive, and they look for behavioral evidence of depression in addition to subject reports about mood. They do not use etiological criteria to classify, for that would also raise issues about circularity and testability. So Brown and Harris's data are not based on a narrow circle of interdependent assumptions nor do they set up their tests in such a way that it is impossible for them to give up their basic hypotheses.

Finally, Brown and Harris try to include the subject's meaning or understanding without making their own results subjective, something Taylor thought inevitable. Brown and Harris propose, among other things, that adverse life events are an immediate cause of depression (and that the frequency of such events is determined by social variables). They realize that life events *as perceived by the subject* are likely to be a key variable. So they use information given by each subject about what the subject perceived to be adverse events preceding the onset of depression. Doing so allows them to include the subject's meaning and to allow that meaning to be contextual, that is, unique to each individual. However, Brown and Harris also think that subjects can misinterpret the events that really influence them – by, for example, looking at every event as connected to their problems to rationalize their situation. So they devise an alternative measure of adverse life events, this one tested on a group of individuals not currently reporting depression. They also control for other factors that on either measure would not count as a major adverse event. The upshot is that Brown and Harris include subject's meaning as a possible factor without assuming that the subject is infallible or making their own account subjective. They use all the steps described above to eliminate bias from their ranking of subject's perceptions, they allow for contextual variables, and they separate common value judgments about what is "adverse" from their measurement of subject's meaning.

Thus Brown and Harris's work provides a prima facie case that interpretive social science can avoid the concerns outlined in Sections 6.2–6.3. It also provides a prima facie case that interpretive social science can provide causal explanations that survive fair tests, cross tests, and independent tests, because the various procedures described above to eliminate bias, to ensure non-circular tests, and so on are concrete embodiments of those testing virtues. Like the work of Paige and of Hannan and Freeman, Brown and Harris's work is not beyond criticism. But arguably it does a relatively good job of meeting basic standards of scientific adequacy. So if critics of interpretive social science want to make a convincing case, they will have to leave behind armchair examples and argue over the specific details of work like that Brown and Harris.

6.5 Norms and symbols

In this last section we look at several specific kinds of interpretive social science that have seemed especially troublesome. In what follows I examine two controversial interpretive approaches – those invoking norms and symbolic meanings – and argue that the problems are not inherent.

I start first with interpretive social science at perhaps its most outrageous – with symbolist anthropology that sees meanings that are meanings for no one. Symbolism is a broad school within anthropology that goes back at least to Durkheim and remains an ongoing research tradition. Its more recent practitioners include Turner (1967), Lévi-Strauss (1967), and Geertz (1973a). Though these individuals differ on specific issues, they share the beliefs (1) that good social science involves finding the symbolic meaning of everyday social practices and (2) that those meanings are not necessarily ones explicitly expressed by individuals. Turner, for example, argues that the milk tree in Ndembu rituals represents the division between mothers and daughters and men and women – despite the fact that the Ndembu themselves see it as representing unity. Geertz sees the Balinese cockfight as a "sustained symbolic structure" or "an ensemble of texts" that "says something of something" (1973a, p. 450); it is a "metasocial commentary upon the whole matter of sorting human beings into fixed hierarchical ranks" (1973a, p. 448). Again his reading goes beyond anything the Balinese explicitly affirm. The work of Lévi-Strauss and many others follows a similar pattern.

Some philosophers worry that the symbolist enterprise borders on a conceptual confusion (Skorupski 1976; Papineau 1978). Symbolists are committed to meanings that are meanings for no one; that basic idea, however, looks incoherent. While we can sympathize with such worries, I do not think the problem with symbolism is primarily a conceptual mistake. Are meanings that are not meanings for anyone impossible? It is doubtful that any conceptual fact about meaning will decide this question, especially given how difficult it has been to develop a theory of meaning. Instead, symbolist theories have to be judged by how well they succeed at the usual scientific tasks. However, "hidden" meanings may have some explanatory value. We invoke unconscious or tacit beliefs all the time in ordinary explanations of behavior. Moreover, appeals to "information" and other cognitive states at the subconscious level are common from cognitive psychology to ethology to neurophysiology. Such accounts may not in the end of inquiry be part of our best science; yet their very presence and pervasiveness suggests that symbolism's problem is not that it rests on a conceptual confusion about meaning.

Symbolism's problems lie elsewhere – in the practical difficulties of providing well-confirmed explanations. If Quine and Rosenberg's objections to interpretive social science are avoidable in principle, they haunt symbolist practice with a vengeance. The basic problem is that we have no very good idea of just what implicit or unconscious belief comes to and thus no very definite set of constraints on its attribution. Thus multiple interpretations are an ever-present worry. Different anthropological approaches are likely to come to quite different results. Disagreements can be across the board and leave little shared data to provide fair tests. The best symbolist work does try to integrate interpretation with what else is known about kin structure, linguistic change, economic relations, and so on. But those constraints are often not enough to decide whether the Balinese cockfight, for example, is best described as a "comment" or given some other meaning. It does not take an elaborate story about indeterminacy or the failures of belief-desire psychology to recognize such things – these problems and others are widely cited in the anthropological literature (Keesing 1987; Thomas, Kronenfeld, and Kronenfeld 1976).

Aside from these problems about confirmation, there are further problems about explanation. In what sense is identifying the meaning of a social practice an explanation of it? Let's pursue this question for a moment on the presumption that explanation revolves around citing the causal facts. Our question then becomes: Do symbolist accounts provide causal explanations? It is conceivable that they could, if the postulated meanings were taken as implicit beliefs that played some causal role in behavior. Yet even when symbolists take their interpretations as about implicit or unconscious beliefs, they seldom give them any clear or believable causal role. Moreover, with the likely possibility of multiple interpretations, we run the risk that no interpretation picks out the real underlying psychological mechanism. Symbolist accounts are by definition "thick" – they make strong assumptions about meaning and we cannot rescue them from the psychological validity problem as we did Brown and Harris's work.

If we take explanation as about unification, the symbolists account may fare somewhat better. Lévi-Strauss's work, for example, does apply a simple pattern to a wide range of diverse social practices; Turner's "forest of symbols" does fit Ndembu rituals into a comprehensive classification system. On my view this fact just illustrates again the problems with taking explanation as essentially about unification – particularly, the problems with specifying and motivating an objective account of unification. Symbolist patterns of meaning look more like interesting regularities that call for explanations rather than explanations themselves.

If symbolism is an example of interpretive excess, that does not mean it must remain so or that it has no value. As a search for unifying patterns, symbolism can be a useful propaedeutic to more solid explanation. And nothing, so far as I can see, rules out the possibility that explanations in terms of implicit beliefs might give well-confirmed causal accounts. To do so, however, will require much more careful accounts of what implicit meanings involve and of their causal role than is currently found in symbolist work.

A second body of interpretive work that is also often alleged to be empirically and explanatorily sterile is that based on normative meanings. "Normative meaning" refers to such things as norms, rules, conventions, roles, and so on. Much sociological and anthropological explanation invokes these normative meanings, though their specific content varies. Exactly how these different phenomena interrelate is a matter of ongoing debate, and usage of these terms is not always uniform. Here I understand them as follows: Rules are prescriptions on behavior, imperatives that say what ought to be done. However, rules can exist without eliciting compliance – there is, for example, a rule against traveling in basketball which is nonetheless frequently ignored in the NBA. Norms, as I use the term, are also imperatives, but they are imperatives that are reflected in behavior. In short, norms, unlike rules, describe real patterns of behavior. Conventions, following Lewis's (1969) usage, are like norms in that they describe patterns of behavior, but they differ in that they are more or less the voluntary agreements resulting from pursuit of self-interest. Roles are then some complex intersection of norms, rules, and conventions.

Skeptics doubt that explanations involving normative meaning can be good social science. I share their doubts, but I deny that the problem is unavoidable. Below I sketch some common criticisms, arguing that the problems they identify are important reminders rather than inevitable flaws.

As they do for meaning in general, interpretivists sometimes claim that understanding normative meaning also involves special methods and provides special explanations outside the pale of natural science standards. However, many norms, rules, conventions, and so on involve explicit beliefs – the beliefs that some behaviors ought not be done, that some behaviors deserve certain sanctions, and the like. Understanding those normative meanings is thus not in principle different from understanding beliefs and linguistic meaning. As we argued earlier, insight and empathy may provide evidence, but they do so only insofar as they provide cross tests, fair tests, and independent tests. So insofar as normative meaning involves beliefs, there is no good reason to think it lies outside naturalistic investigation.

What about normative meanings that do not involve explicit beliefs – for example, rules of behavior that no subject can explicitly state? Such rules might seem to require some special investigative methods, most obviously "participating" in the culture in question. When such participation allows the social scientist to "act like a native" – to know how to follow the rules – the social scientist has then grasped the rule. However, if the social scientist ends up with only the same tacit understanding as do the subjects, no discursive knowledge has been produced. If participating allows the social scientist to formulate a rule explicitly, then he or she needs to cite the evidence that participating provided. Hence ordinary scientific virtues are still decisive.

Do normative meanings provide some special kind of *explanation,* even if they are confirmed by ordinary routes? Or, to phrase the question as would the right-wing critics, do they explain at all? Since normative meanings involve attributing beliefs, they can in principle provide causal explanations. Consider, for example, using norms or rules to explain. In this case we explain by citing beliefs that some specific behavior is proscribed and then by showing that those beliefs are partial causes of the behavior in question, perhaps via sanctions or socialization. Tracing out these processes is giving a causal explanation. We can thus construe normative explanations as causal explanations and thus as compatible with naturalism.

Nonetheless, there are serious potential obstacles to explaining via norms and rules. For starters, it is not at all obvious that simply *describing* the norms or rules of a society is an explanation at all. Suppose we describe common patterns of behavior or we depict the beliefs, explicit or implicit, involved in a norm without tying those beliefs causally to behavior. If explanation is about citing causes, we have not explained (a point made earlier about symbolism). Like early zoologists who did taxonomy without the use of evolutionary theory or chemists who categorized compounds without atomic theory, social scientists who describe belief systems without tying those beliefs to causal explanations are arguably not explaining at all. Of course, we have good reason to think that attributing beliefs and providing causal explanations go hand in hand – and, as we pointed out earlier, much supposed interpretivist research is implicitly or explicitly about causes. However, if explaining via norms simply describes beliefs – and if we think causation is central to explanation – then appeals to normative meanings are not explanations. That, of course, does not mean they are not valuable for other purposes.

Normative meanings face other problems. Rosenberg's multiple models problem is likely. Though multiple models need not preclude explanation, they do cause real problems if we make specific assumptions about mechanisms. However, explanations by appeal to norms – interpreted as causal explanations – claim that the norm in question is the causal mechanism behind behavior, so the concern is real. Furthermore, non-causal accounts presume there is one best way to unify, one best way to describe the rules of given society, and so on.[12] Thus if alternative hypotheses about norms have equal empirical support, we have no reason to claim any particular normative explanation is the right one. At most, the different alternatives would give us different ways of systematizing behavior – but no explanation. However, showing this potential problem to be a real one requires making specific arguments about specific alternatives to specific proposed normative explanations. I am sympathetic to the idea that much normative explanation may be no such thing for the reasons cited here; I see no reason, however, to think every normative explanation must fail.

Another concern about normative explanations is that they are not "deep" enough. When sociologists use norms to explain, advocates of rational choice theory, for example, are not satisfied: Why then, they ask, is there such a norm? Some go so far as to claim that a rational choice explanation in terms of maximized self-interest always trumps an appeal to norms, and that appeals to norms are always insufficiently "deep" (Hechter 1983).

This concern nicely illustrates some points made in Chapter 2 about scientific virtues. I argued there (1) that claims about good science are contingent empirical claims and (2) that abstract virtues play a role in scientific disputes by being "embodied" in much more concrete traits specific to disciplines. The debate over rational choice versus normative explanation illustrates both these points. If self-interested rational choice explanations always trump other explanations, it is not because of some conceptual fact about human behavior. Rather, this claim could only be warranted as an empirical generalization: given what we know about the history of past explanatory successes and failures, good explanations will have such and such a form. Similarly, claims that we need deeper explanations are claims about how much of the causal network we need for explanation. Aside from pragmatic interests, our background knowledge will determine when explanations are "shallow" and "deep." For example, the rational choice assumption is that social practices must on the whole be compatible with individual self-interest. When altruistic behavior is ex-

---

[12] See P. Roth (1987, ch. 6) for the extent to which Winch relies on the assumption that there is one set of rules to uncover.

plained by appeal to norms, that explanation has not yet rendered the behavior consistent with the rational choice constraint on explanation – thus it is not sufficiently deep.

So normative explanations are empty only to the extent that the rational choice strictures on explanation or others like them are well confirmed, an issue I will not try to settle here. In all likelihood, no single answer will be forthcoming. For some domains of behavior, the rational choice demand may be reasonable. Yet in others appeals to norms along with socialization mechanisms may be enough, for the assumption of self-interested behavior may be less convincing.

I want to finish this chapter by taking up one last doubt about normative accounts, a doubt that we could have raised about symbolism as well. Explanations involving norms and symbols – or, more generally, "culture" – may seem ontologically extravagant. One easily gets the picture of some supraindividual entity working behind the backs of individuals. No doubt social scientists sometimes talk this way. Yet our discussion of individualism in the last chapter shows that these explanations need not commit ontological sins. As interpreted here, norms, social meanings, conventions, and their ilk are realized in complex states of individual beliefs, dispositions, and relationships. Norms, in the language of Chapter 5, are exhausted by facts about individuals and supervene upon them.

We can deflate norms in this way while nonetheless denying that normative explanations are reducible and eliminable. We saw in the last chapter that exhaustion and supervenience do not *entail* reducibility, and problems like multiple realizations and presupposing social information can make reduction impossible in practice. These latter obstacles seem particularly likely when it comes to norms, social meanings, rules, and conventions. The specific individual beliefs and dispositions that realize a norm are likely to vary (1) from individual to individual, (2) from context to context, and (3) from society to society. Similarly, explaining norms in terms of individual behavior is also likely to invoke social roles, group membership, and other social features, thus not eliminating but presupposing social accounts. I will not argue for these claims in detail here, but they indicate that there is a legitimate sense in which norms are "emergent" – a sense which does not commit us to a mysterious ontology.

So there is a strong prima facie case that neither normative nor symbolist explanation is inevitably flawed, though flaws in practice are no doubt plentiful. Interpretive social science is not social science at its best, and normative and symbolist work is not interpretive social science at its best either. Yet even here there are no conceptual or even ineliminable practical obstacles to doing good science. Thus interpretive social science provides no compelling evidence against the main theses of this book: we have seen that there is good interpretive work as illustrated by Brown and Harris, that even the most suspicious interpretive social science is not inevitably flawed, and that even if all interpretive social science was doomed to failure, the prospects for good social science at the macrosociological level would remain untarnished.

7

Economics: a test case

Economics presents an important test case for the claims of this book. Modern economics raises doubts about naturalism because of both its failures and its successes. Its successes come from its sophisticated use of mathematics, making it the social science closest in form to natural sciences like physics. Yet to many, economics seems a dubious science at best, for its models are based on highly unrealistic assumptions and *a priori* theorizing. So if the best of the social sciences is suspect, naturalism is threatened. Economics also challenges the holism defended in Chapter 5. Methodological individualism is the official philosophy of mainstream economics. Its most powerful models seem to work by describing laws of individual behavior and deducing aggregate social phenomena from them. Here individualism appears to get its strongest support.

This chapter makes a prima facie case that economics is no threat to either holism or naturalism. The discussion proceeds as follows. Section 7.1 sets the background. I outline some standard views about neo-classical economics and discuss their specific inadequacies. Most of these views err in treating modern economics as a seamless whole that stands or falls together. Section 7.2 rejects that assumption, arguing that the laws of *aggregate* market supply and demand are relatively well confirmed – and are independent of the highly unrealistic models of abstract equilibrium theory. Section 7.2 also argues that the laws of supply and demand constitute a central "argument strategy" in economics that undergirds much economic explanation and that it is an error to treat the highly idealized neo-classical models as the whole of modern economics. Section 7.3 takes up the status of those neo-classical models. Section 7.4 turns to issues concerning reduction and individualism.

7.1 How to think about economics

Philosophers and economists have written extensively about the scientific standing of economics. They have done so, of course, because economics is so puzzling: economics has highly developed mathematicized theories and yet it is not obvious that those theories explain any real economy at all, because they make such violent abstractions from reality. In this section I survey several attempts at resolving this puzzle. Doing so will help set up the issues for the following sections and help further flesh out the naturalism defended throughout this book.

Let's first recall what basic models in modern economics look like, so that we are clear about just how puzzling its scientific status is. At least some or all of the following are typical postulates in modern economic theory:

(1) Individuals have a complete set of preferences over all goods and those preferences are transitive.
(2) Individuals know the prices of all goods.
(3) Firms know the prices of all goods.
(4) Firms know the most efficient technological relation between inputs and output at a given price.
(5) There is a market for every good at every time.
(6) Firms maximize profits.
(7) Individuals maximize utility given their resources.
(8) There is complete mobility of labor and capital across markets.
(9) Individuals and firms carry out all agreements.
(10) There are sufficient mechanisms, time, and so on to ensure that markets can reach equilibrium.

These postulates obviously leave no or little room for mistakes, limited information, non-economic causes, groping towards acceptable solutions, incomplete markets, cheating on contracts, collusion, continuous and rapidly changing economic environments, and so on. Since these latter conditions are typical of real economies, the puzzle is how theories based on postulates like those listed above could explain real economies and how facts about real economies could count as evidence for theories embodying assumptions like these.

The *locus classicus* for recent discussion of the scientific standing of economics is Milton Friedman's (1953b) "The Methodology of Positive Economics." That essay was written at a time when critics of neo-classical economics were arguing that real firms do not actually attempt to equate marginal cost and marginal revenue, key assumptions of the neo-classical theory of the firm. For Friedman, these criticisms are irrelevant. Predictive success – along with simplicity, consistency, fruitfulness, and the like – determines which theories are adequate and which are not. There is no testing a theory by its assumptions, only deriving implications from a theory and comparing them with reality. Friedman also argued that all theories make unrealistic assumptions. Thus irrealism is not the issue. Assumptions are "realistic enough" if they yield sufficiently accurate predictions; which predictions are "sufficiently accurate" is a pragmatic matter, depending on the purposes at hand. When a model or theory entails sufficiently accurate predictions, we are justified in claiming that the world acts "as if" the theory is true. That much is all any science can claim.

How then does economics fare according to Friedman? It does relatively well. Price theory and static monetary theory are relatively well confirmed. Where weaknesses do exist – as in dynamic monetary theory – they do not result from unrealistic assumptions, but from failure to generate implications from abstract theories. Friedman seemed to give economists free reign to develop abstract models with wildly false assumptions. They have not hesitated to follow his lead.

Friedman's argument seems to assume that only predictions count. That instrumentalist approach, of course, came under heavy criticism with the demise of positivism. More recent work on economics has thus tried to apply post-positivist insights. For example, Weintraub (1985) has argued that Lakatos's conception of a scientific research program sheds considerable light on economics.[1] For Lakatos, a scientific research program includes at least: (1) a "hard core" – a defining set of unrevisable assumptions, (2) a "protective belt" of auxiliary assumptions and more specific theories that instantiate the hard core and make testing possible, (3) a positive heuristic that specifies how to revise the protective belt in the face of anomalies, and (4) a negative heuristic that enjoins revising the hard core. For Lakatos, only research programs as a whole could be evaluated. They were either progressive or degenerating: degenerating if theories in the protective belt are developed solely as ad hoc responses to falsification, progressive if they successfully predicted novel facts.

Weintraub believes that it is essential to distinguish between the hard core of the neo-classical paradigm and its subsidiary theories in the protective belt. General equilibrium theory constitutes the hard core. Roughly speaking, general equilibrium theory begins with Walrus, is developed by Cassel (1932), von Neumann (1945), and Arrow and-Debreu (1954), and is perhaps most clearly stated in Arrow and Hahn (1971). That tradition

---

[1] Weintraub has since abandoned this general approach for social constructivism. See Weintraub (1991).

works, Weintraub claims, by assuming that agents maximize and have full preferences and knowledge and that all explanations must refer to equilibrium states. These assumptions constitute the hard core and are not up for debate. Human capital theory, monetarist theory, optimal tariff theory, and so on constitute the protective belt.

Weintraub thus ultimately provides yet another defense of economics. Unrealistic assumptions are part of the hard core and not what we should evaluate. The general equilibrium research program is progressive if it leads to successful theories in the protective belt. Weintraub believes that it has and thus that the general equilibrium research program is progressive.

Weintraub defends neo-classical theory by recognizing that testing is more complicated than the positivists realized. Donald McCloskey (1985) defends economics by a similar strategy, though his rejection of positivism is considerably more radical than Lakatos's. According to McCloskey, neo-classical economics has an official methodology that is at variance with its practice. Its professed methodology is "modernism," which for McCloskey is roughly the picture of science and scientific methodology described by the positivists. Modernism is thus the idea that science proceeds by applying a set of universal, *a priori* standards that, along with objective data, make science a rational and objective enterprise. While this is the professed method of economics, actual economic explanation employs quite different methods. Empirical evidence plays a minor role at best; instead, numerous rhetorical devices are what really persuade.

McCloskey does not, however, see cause for pessimism in all this. Rather, he argues – or, more accurately, asserts – that the positivist ideal was one which no science can achieve and that all science is a matter of what is socially persuasive. Thus McCloskey adopts a view of science much like that of Rorty. Good method, on his view, is that which persuades in economic conversation.

While McCloskey's views may irritate some economists, he has in fact provided yet another defense of the discipline. Economics is doing relatively well, for it has a rich and creative set of persuasive practices. Since no science produces objective data, economists need not worry about charges of empirical irrelevance and the like. Economics can no doubt be improved, but McCloskey seems quite clear that the way to progress is not through more and more carefully designed tests and more realistic models. His radical stance on method thus disguises a very conservative stance on substance.[2]

[2] Which probably explains why McCloskey's views have gotten good press in conservative publications like *National Review* that are otherwise appalled at postmodernist excesses. See Ryan (1991).

It is not surprising to find economists like Friedman, Weintraub, and McCloskey defending their enterprise. It is also not surprising that philosophers have been more skeptical. I want now to look at the views of two such philosophers, Alexander Rosenberg and Daniel Hausman. Rosenberg has written extensively and insightfully about economics for nearly two decades (1976, 1981, 1983, 1989, 1992). In his earliest work, *Microeconomic Laws* (1976), Rosenberg argued that there are no conceptual obstacles to economic science. Microeconomics produces statements with the right form to be laws; although those statements explain economic behavior by appeal to reasons and beliefs, those statements can nonetheless cite true causes, for reasons can be causes. *Microeconomic Laws* did not argue that modern economics is empirically successful, but it denied that there were inherent obstacles to its being so.

Rosenberg later became much more circumspect. He argued that there has been little empirical progress in economics since Adam Smith. Economics at best produces generic predictions – predictions of rough causal directions – but generic predictions are not good enough, at least if left unrefined and unrefinable. Moreover, neo-classical economic theory has consistently been falsified and revised ad hoc. For example, the original cardinal utility accounts of individual choice were replaced by ordinalist notions and then those by revealed preference. In each case empirical inadequacy, not theoretical enrichment, motivated the change.

The real puzzle is thus why economics has failed and how we should think about the obviously creative work that economists do. Economics has failed, Rosenberg argues, because its belief-desire psychology is inherently flawed. We saw Rosenberg's reasoning in the last chapter: belief and desire constitute a closed circle that cannot be broken; economics thus has no prospects for refinement. So what is it that economists are doing? Rosenberg entertains two ideas. We can think of economics as a branch of applied math – as developing the implications of imaginary models or of axioms defining rationality. Alternatively, we might see in neo-classical economics a social contract argument for the market. Proofs that equilibrium prices exist and are Pareto optimal are proofs that the market system can efficiently reconcile competing interests through voluntary exchange, thus providing a convincing contractarian argument for the market.

Hausman reaches rather different conclusions (1981a, 1981b, 1984, 1992). He makes five basic claims. (1) The fundamental theory in economics is "equilibrium theory," which assumes complete and transitive preferences, non-satiation, profit maximization, diminishing returns, and the existence of equilibrium prices. General equilibrium theories (GETs) – often thought to be the most fundamental version of modern neo-classical economics – are then augmentations of equilibrium theory and result by add-

Economics: a test case 227

ing assumptions such as perfect knowledge, complete interdependence of markets, existence of futures markets, and so on. (2) The basic laws of equilibrium theory are taken by economists to make an "inexact and separate science" – inexact in that they must be qualified *ceteris paribus* and separate in that these laws describe "a peculiarly economic realm," a small set of specifically economic processes that account for most economic phenomena. (3) Equilibrium theory is tested with mixed success by economists, due to the recalcitrance of the subject material, not methodological errors. The basic laws of equilibrium theory are *a priori* plausible, while the quality of economic data and the many auxiliary assumptions needed for econometric tests are generally quite questionable. As a result, economists rightly do not reject equilibrium theory upon predictive failure. (4) Economists are dogmatic, however, when they downplay predictive failures by arguing that other approaches do not treat economics as a separate science. Whether a separate science is possible is an empirical matter, and not a claim that can be justified *a priori*. (5) General equilibrium theories – the most idealized and abstract form of modern economics – are neither well confirmed nor explanatory, because they rest on extremely unrealistic assumptions that do not meet the kind of requirements on *ceteris paribus* clauses that we outlined in Chapter 3.

Though these five accounts all can teach us something about economics, none is fully adequate. Some failings are particular to the individual views, others rest on common assumptions. I spell out some of these problems immediately below; other criticisms will emerge as the chapter progresses.

Milton Friedman's (1953b) article has been discussed – and criticized – at length. It will be useful, however, to summarize its problems, particularly as they relate to earlier discussions:

(1) Treating the assumptions or basic axioms of neo-classical theory as mere "as if" accounts makes them non-explanatory. Whether we treat explanation as unification or causation, we will want the explanans to be *true*. But that is precisely what the "as if" reading gives up. So Friedman's defense leaves economic theory unable to explain.
(2) We cannot easily segregate assumptions from their implications. Every statement entails itself. Thus one of the implications of neo-classical theory is that its axioms hold. So any theory with unrealistic axioms makes unrealistic predictions as a matter of logic.
(3) Friedman is right that unrealistic assumptions are no inherent obstacle – that was a moral we drew in Chapter 3. But that does not mean anything goes. Unrealistic assumptions are allowable

only if they pick out real causal factors. And to show real causal factors we need more than accurate predictions: we also need to rule out spurious successes and competing accounts, as we saw in Chapter 3. Without these steps, we are left with a weakly confirmed and unexplanatory theory.[3]

These criticisms are devastating. However, there is a core insight in Friedman that is worth emphasizing: simply pointing to unrealistic assumptions is not enough; since unrealistic assumptions are inevitable, only a careful assessment of the empirical details will tell us whether neoclassical theory succeeds. And to be fair, this insight was probably Friedman's main point, a point lost in the other confusions of his article and the use to which his arguments have been put. Hirsch and de Marchi (1990), for example, argue convincingly that Friedman's main concern was to combat the *a priori,* introspective evidence and theorizing so common inside economics. Friedman emphasized predictive accuracy as an antidote to these trends, not as a blanket defense of the neo-classical approach. In fact, Friedman's own approach to empirical work often reveals the testing virtues identified in Chapter 2. His research practice and his official statements on methodology are not entirely consistent.

Weintraub's Lakatosian defense both inherits the inadequacy of Lakatos's philosophy of science and begs the question. Lakatos's views are generally an improvement over positivism, but they suffer from at least two problems. First, hard cores *are* tested. For example, "no action at a distance" was central to the mechanistic program of early modern physics; it was eventually given up because of Newton's striking empirical success with laws that allowed action at a distance. We should not be surprised by this problem, for Lakatos's shielding of the hard core really amounts to smuggling the analytic–synthetic distinction back into philosophy of science. Second, predicting novel facts is not the sole or overriding criterion of empirical support. Philosophers of science have reached no consensus over whether new facts confirm more than old ones; a reasonable guess is

---

[3] Friedman at points justifies his approach by arguing that the assumptions of neoclassical theory are unobservable – that they can only be tested indirectly. Yet that is doubtful as a general statement. If we take profit maximization, for example, as one of the assumptions of the theory Friedman is defending, then it surely is relatively directly testable – by evaluating firm costs and pricing behavior. What is harder to observe is individual actors consciously seeking to maximize profits. Yet that is not the relevant assumption for most situations, but rather the assumption that firms do, by whatever mechanism, maximize. That assumption can be tested on its own, independently of embedding it in some larger theory and deriving implications.

that there is no *a priori* answer – the importance of new versus old depends on the substantive empirical context. There are good reasons to think that *when* the data were discovered is not really the issue. Rather, the key question is whether our background information suggests that the data are improbable on every other competing explanation.[4] In short, new data carry special weight only when they make for stronger fair tests that rule out competing explanations. Whatever the outcome of this debate, predicting new facts will still be just one of several scientific virtues. A theory that generates more novel predictions than its competitors but that does so by introducing many new false or unsupported claims may be inferior, not the most progressive.

These two problems undermine Weintraub's defense of general equilibrium theory. The unrealistic assumptions of neo-classical economics are not immune from empirical test. It may take serious empirical difficulties to warrant abandoning them, but it is not a conceptual confusion, as Weintraub maintains, to ask about their empirical justification. Secondly, the progress Weintraub cites – the development of theories in the protective belt such as human capital theory, optimal tariff theory, and so on – does not necessarily show the general equilibrium paradigm to be progressive. Novel predictions may come at the cost of decreased realism, for example. Weintraub needs to show that these theories are improvements when other evidential virtues are included, something he does not do.

Moreover, Weintraub has not shown economics to be progressive, even in Lakatos's own, flawed sense. He largely just asserts that there has been progress in the protective belt.[5] But that claim assumes, for example, that human capital theory is predicting novel facts that are empirically confirmed. However, if that was uncontroversial, there would be much less worry about the status of current economics – and much less worry about economic idealizations. Maybe Weintraub is right.[6] But the real task is to establish such progress, not to assume it. Moreover, economists like Hands (1993) who have looked in detail for signs of Lakatosian progressiveness fail to find it. Thus while Weintraub's defense is provocative and interesting, it skirts the real issue.

Chapter 2 has already laid the ground work for evaluating McCloskey's rhetorical approach. Like Rorty, McCloskey assumes that there are only two alternatives: a crude positivist methodology or the equation of good

---

[4] For discussions of this issue *roughly* along these lines see Howson and Urbach (1989) and Maher (1988).
[5] Though no evidence is provided in Weintraub (1985), he does provide some support for this assumption in Weintraub (1988).
[6] The empirical evidence on human capital theory is far from conclusive, or so Blaug (1976) argues.

methodology with what persuades. McCloskey (1985, p. 7) sees the official methodology of economics as advocating a timeless set of *a priori* methodological rules that sharply separate science from non-science, deny that there can be non-experimental evidence, take falsification to be the essence of testing, assume a sharp separation between theory and observation, think of testing as a logical relation between a hypothesis and the data alone, and so on. Perhaps this is the official methodology of economics. However, McCloskey's science as persuasion is not the only alternative. Chapter 2 defended a picture of method which denies every one of these positivist assumptions and yet which does not make good science simply a matter of persuasion. We can allow non-experimental evidence, the theory-ladenness of observation, the importance of background assumptions, and the contingent nature of methodology, and nonetheless still believe that there is rational, objective empirical warrant for some theories over others. We can do so by providing fair tests, cross tests, and independent tests – of theories, assumptions, and methodological norms. Opting for "methodology is persuasion" is taking the easy way out and ignoring the real prospects for an economic science.

Nonetheless, McCloskey's rhetorical approach, if stripped of its subjectivist ideology, may provide important insights. Looking at how economists persuade can tell us much about how well the field is doing and where it might be improved. We want to know if economists persuade by *reliable* means – ones that on the whole are likely to lead to the truth. Understood this way, analyzing economic rhetoric is simply part of the naturalistic approach. Diagnosing economic rhetoric can make very useful contributions, taken in this light.[7]

Let me turn now from the defenders of modern economics to its critics. I sympathize with Rosenberg's worries but do not share his pessimistic conclusions. He thinks it obvious that economics has failed; the major question for him is why it has done so. However, I shall argue below that Rosenberg's obvious assumption is wrong: important parts of modern economics are relatively well confirmed. Moreover, where modern economics fails – and those areas are nonetheless large – it does not do so because of the vicissitudes of belief-desire psychology. We argued in the last chapter that there are numerous concrete methods for dealing with the belief-desire circle. Moreover, I shall argue that important parts of economics can proceed without invoking the belief-desire psychological

[7] For example, work like Mirowski's (1989) attempt to track down the influence of physics metaphors on neo-classical economics can contribute much to understanding what a better economics might look like.

assumptions Rosenberg finds so troubling. So if economics fails, the explanation lies elsewhere.

Aside from their specific problems, the approaches to economics discussed above also face some common problems. For one, they all – with the exception of Hausman's – tend to treat economics as a seamless whole that we can evaluate in one fell swoop. For Weintraub, economics is a progressive research tradition centered on general equilibrium analysis. McCloskey defends the rhetoric of economics. Rosenberg denies that economics has progressed much since Adam Smith. In each case, the object of evaluation is a very large body of work, nearly the entire field. This monolithic approach is a mistake. While hypotheses are not tested in isolation, it does not follow – as we saw in Chapter 2 – that everything stands or falls together. Real assessment requires attention to real evidence. The web of belief has strands, some of which are more tightly interconnected than others. "Economics" or even "neo-classical economics" denotes an enormous variety of work involving different assumptions, methods, quality of evidence, and so on. Blanket criticisms or defenses are unlikely to be very helpful.

This monolithic treatment raises a second problem. The views discussed above equate modern economics with a particular kind of theory, namely, the abstract neo-classical models grounded on assumptions of profit maximization, decreasing marginal utility, well-ordered preferences, complete markets, and the like. Yet one of our morals from Kuhn was that scientific practice is a complex structure, one that often goes beyond the official models reported in textbooks. Thus any account of what "defines" a given science requires careful analysis and evidence. The commentators discussed above do not provide such evidence, though Weintraub and Hausman make useful first starts. On the whole, commentators simply assume that the highly unrealistic models of equilibrium theory exhaust or best represent modern economics.

That assumption is unwarranted. Abstract models play many roles in economics, but they are only part of the field. Modern economics, I shall argue, fits the model of social explanation defended in this book: explanation comes from applying causal generalizations that specify some of the main variables and then factoring in causes unique to the case at hand. In economics, the laws of supply and demand provide one such set of generalizations. A fundamental explanatory strategy in economics is to apply those basic laws and the causes they identify to more and more specific situations, adding in local complexities. These causal generalizations, I shall argue, do not presuppose the abstract models of equilibrium theory and do not inherit their problems.

I sketch this alternative picture now without argument to drive home my objection to previous accounts: just what constitutes modern economics and just what role abstract models play in that enterprise are complicated questions that must be carefully addressed. Commentators who equate economics with abstract model building are making an unargued-for assumption, one that can sharply influence subsequent evaluation of the field.

Thus the status of modern economics remains wide open. Assessing economics requires looking at economic practice piece by piece and asking whether it meets the broad traits of good science outlined in Chapter 2. The next two sections make some forays in that direction, albeit tentative and partial ones. I hope at least to make a prima facie case that there can be and sometimes is good science in economics.

7.2 The supply-and-demand core

In this section I defend the idea that economics can and sometimes does have the virtues of good science by arguing: (1) that the laws of supply and demand are relatively well confirmed for a range of markets, (2) that those laws can be confirmed independently of the unrealistic models of general equilibrium theories (GETs) and their cognates, and (3) that the laws of supply and demand offer an argument pattern of great significance in economics. Along the way I discuss the role of modeling in economics, arguing that abstract models like GETs have a different and less important role from that pictured by most commentators.

7.2.1 Confirming the laws of supply and demand

The "laws of supply and demand" have referred to numerous different ideas in the history of economics. They are now often associated with the neo-classical theory of consumer behavior. My interest, however, is in a different notion. At their most basic level, I take the laws of supply and demand to claim that price, aggregate demand, and aggregate supply are causally interdependent. Increases in price decrease demand and increase supply; increases in demand and decreases in supply raise price. More subtly, changes in demand and supply can be subdivided along two dimensions. Shifts in the demand or supply curve – that is, changes in the quantity demanded or supplied for each price – are different from changes in the quantity demanded or supplied – movements along the curves – due to price changes. Shifts in the curve influence price; movements along the curve result from changes in price. These two kinds of processes together produce the market mechanism that tends to balance supply and demand. What are usually called the laws of supply and demand are claims about movements along the relevant curves where price is the independent variable. Nonetheless the full story about supply and demand includes both kinds of changes. In empirical work both processes must at least implicitly be considered to infer causes from correlations between price and demand or supply.

In what follows I cite a systematic body of research confirming the claims listed above. It is important to note several things about these supply-and-demand explanations. First, the work I present concerns *aggregate* (or market) demand; it is about the large-scale processes of markets. What it says about individual behavior is an open question I address later. Second, the work I discuss produces laws only of the sort I have defended throughout the book, namely, causal generalizations that have to be applied case by case, factoring in other complicating causes. In short, these supply-and-demand explanations identify tendencies or partial causal factors.

Let's look first at the evidence for the law of demand. Economists have worked long and hard to confirm the law of demand. To those non-economists who have taken formal courses in economics, this may come as a surprise. Economics textbooks seldom mention this work. Unlike in physics where important empirical results get frequent mention, economists generally support the law of demand by deriving it from GET-like assumptions about constrained choice. It may even be that many economists believe it for this reason. Yet they need not do so – and probably should not, for better evidence is available.

Testing the law of demand at first sight may seem straightforward. The law claims that changes in quantity demanded should be a function of price. Thus we should use statistical information about markets to estimate the equation:

$$Q_x = e_x P_x \tag{1}$$

where Q is the quantity of good X demanded, P is its price, and e is a regression coefficient measuring the relative influence of price on demand or, in other words, the elasticity of demand. If the law of demand holds, then e should be negative.

Of course things are not this simple. The influence of price on demand should depend on how much income consumers have at their disposal. So aggregate income in the relevant market will have to be controlled for by adding in income variable to equation (1):

$$Q_x = e_x P_x + e_i I \tag{2}$$

where I is a measure of aggregate income and e_i measures the expenditure elasticity. Nonetheless, equation (2) does not fully handle the effect of income. When the price of a good changes relative to the prices of all other goods, that change means an effective change in income. If the price falls, for example, my real income increases because I can now buy the same quantity of X for less. As a result, we must modify equation (2) to compensate for the income effects of a price change. In particular, we must change the term $e_x P_x$ so that it reflects only the effect of price and not that of income changes; the latter must be subtracted from the elasticity measured. This can be done via the Slutsky equation:

$$\left|\frac{\partial Q_x}{\partial P_x}\right| = \left.\frac{\partial X_x}{\partial P_x}\right|_{\Delta I = 0} - Q_x \left.\frac{\partial Q_x}{\partial I}\right|_{\Delta P_x = 0} \tag{3}$$

which says that the rate of change in quantity of X demanded relative to the rate of price change ($\partial Q_x/\partial P_x$) is a function of that same quantity with income held constant minus the income effect ($Q_x \, \partial Q_x/\partial I_x$). The term $\partial Q_x/\partial P_x |\Delta I = 0$ is called the compensated elasticity and is the value we need to test the law of demand.

Equation (1) also ignores another obvious factor: the price of other goods. The price of goods which can be substituted for X or which are complements to X will influence Q_x as well. So we need to include them in our test of the law of demand. Changes in their prices will also have income effects, so the relevant variables to include are their compensated elasticities. Changes in the price of goods which are neither substitutes nor complements are also relevant to the demand for X because they influence available income. Thus we must control for these changes as well. A natural way to do so is to divide aggregate income by the overall price index, so that changes in the relative prices of other goods will show up in real income. Putting all these complications together, testing the law of demand for a particular commodity thus requires estimating something like:

$$Q_x = e^*_{xx} P_x + e^*_{xs} P_s + e^{**}_{xc} P_c + eI/P \tag{4}$$

where e^*_{xx}, e^*_{xs}, and e^*_{xc} are the compensated elasticity of demand for X (commonly called the "own-price" elasticity) and the compensated "cross-price" elasticities for close substitutes and complements, respectively, and I/P is the income variable in real terms. Equation (4) now captures the basic variables that must be measured to test the law of demand.

Economists began using statistical data to test the law of demand in the early 1900s (Moore 1914, Schultz 1938). However, extensive empirical work really began after World War II, beginning in particular with Stone's (1954) classic studies. Stone separately estimated demand equations for forty-eight goods using aggregate British time-series data. His equations were basically the same as equation (4) above, though Stone also included a further complicating variable: changes in taste. Changes in demand over time might not reflect the influence of price change but rather changes in aggregate preferences. To control for this possibility Stone added a time variable. If market preferences are changing over time, then we should find a correlation between the time variable and the quantity demanded. If changes in preferences entirely explain changes in demand, then adding a time variable should remove the correlation between demand and price.

Stone's results clearly confirm the law of demand. For forty-six of the forty-eight commodities, the own-price elasticities – the measure of how the price of a good directly influences its demand – are negative; the two that are not are also not statistically significant. These results hold up when changes in income, income effects, the prices of other goods, and potential changes in taste are considered. Stone thus shows that many of the *ceteris paribus* assumptions implicit in the law of demand do not undermine his evidence. This evidence is of course far more compelling than the usual anecdotal evidence cited in textbooks.

Developments since Stone have replicated his results and refined both the testing and specification of the law of demand. Let me mention briefly some of those developments:

(1) Barten (1967) replicated Stone's results in the late 1960s using the same single-equation approach but with slight differences in specification of the demand equation and an entirely different data set, namely, Dutch aggregate data from before and after World War II.
(2) In the late 1960s through the mid-1970s, Theil (1975, 1976), Deaton (1974), and Barten and Geyskens (1975) independently estimated simultaneous equation systems on German, Dutch, and British data (the latter being different from the data used by Stone).[8] All the tests show negative own-price elasticities for nearly all goods.[9]
(3) In some of the most recent work, Pollak and Wales (1992) have extended these tests in two different directions. They have estimated consumer demand on data broken down by household variables rather than aggregate market data as in the above work.

---

[8] The best overview of more recent work is probably Deaton and Muellbauer (1980).
[9] These simultaneous equation approaches sometimes impose cross-equation restrictions derived from the theory of individual utility maximization, even though the consumer data is about aggregates. This procedure is suspicious – see Deaton and Muellbauer (1980) and Gilbert (1991). However, the single equation approach of Stone and others does not generally impose such assumptions, and tests of the law of demand using the simultaneous equation approach, but without imposition of constraints from consumer theory, still show that own-price elasticities are negative (see Deaton and Muellbauer 1980, ch. 3). In short, these tests of the law as an aggregate law do not essentially presuppose a neo-classical theory of individual consumer behavior.

They have also investigated other potentially confounding variables such as family size and past consumption. Family size might influence demand in fairly obvious ways; past consumption might affect demand either by habit formation or by a sort of "contagion effect" like that discussed in Chapter 3 where the presence of a behavior spreads through a population. Pollak and Wales test many different specifications of the demand equations incorporating these variables. They find that own-price effects are negative across specifications and remain negative when these potentially complicating variables are included; the relative size of the own-price elasticities is likewise largely preserved across different specifications.

(4) The demand curves discussed here treat prices as exogenous causal factors determining the quantity demanded. This assumes that shifts in the supply curve lead to new equilibrium prices that determine demand; the price-demand data used to test our equations are presumed to be measures of those equilibrium prices and quantities. Both the assumption of exogeneity and that of equilibrium are simplifications of a complex process where price, demand, and supply mutually interact and are not always in equilibrium. Thus it is possible that both assumptions might confound our results and that the own-price coefficients are not accurately measuring the effect of price on demand. Economists (cf. Deaton 1972) typically give informal reasons why these assumptions are reasonable or harmless. However, recent work turns those informal arguments into more rigorous data. Bronsard and Salvas-Bronsard (1984) have investigated whether prices can reasonably be taken as exogenous in demand analysis. They have tested traditional demand models with exogenous prices against more complex dynamic models that make prices endogenous. They find that the law of demand holds up on both models and that estimates of own-price elasticities change very little (both in magnitude and in statistical significance) when endogenous price variables are added. Equilibrium assumptions have also been the target of extensive investigation. The first task is to model disequilibrium and then to find appropriate econometric tools to test those models. Work by Quandt and others towards these goals is starting to influence empirical research, and that research again suggests that modeling demand more realistically by allowing disequilibrium leaves the law of demand intact.[10]

---

[10] For econometric theory on these issues see Goldfield and Quandt (1981) and Quandt (1988). For an example of empirical work, see Yang (1987).

These four developments are just a part of the extensive effort since Stone that have gone into refining and testing the law of demand.

So the law of demand – as a law describing the aggregate behavior of markets – is reasonably well confirmed. No doubt problems exist, but they are much like those faced by Paige, Hannan and Freeman, and evolutionary biology and ecology. In short, at least this part of economics is no threat to naturalism.

Let me turn now to the laws of supply. Like the laws of demand, they claim that (1) when the factors underlying the supply curve are fixed, price causally determines the quantity supplied; and (2) shifts in the supply curve determine price, assuming the demand curve is fixed. And like the demand laws, the supply laws describe the aggregate behavior of markets. The markets I want to focus on most directly are markets for agricultural goods. It is here that the empirical work is probably most advanced and the evidence most convincing.

Literally hundreds of studies have been done on agricultural supply and changes in prices.[11] These studies, like those cited above on demand, are more or less direct tests of the claim that price changes cause supply changes. Supply data for a particular commodity is regressed on price data, controlling for confounding causes. These tests share a basic form. Yet they differ in the control variables, measurement tools, functional forms, and kinds of commodities and data sets used, and in the country, culture, and level of economic development present. This diversity helps ensure that we have cross tests and independent tests, because background assumptions vary widely. While not every study fully supports the law of supply, the vast majority do: they find a positive correlation between price and supply while controlling for other factors.

This work on agricultural supply also does a good job of handling the *ceteris paribus* problem. It controls for numerous confounding influences and has done so with increasing sophistication. Changing technology and factor prices – which would shift supply curves and confound inferences – are controlled for by adding in the prices of inputs to agricultural production to see if the correlation between price and supply vanishes. Technological change is often handled much like potential changes in consumer tastes in demand theory, namely, by using a time trend variable to look for spurious correlations. Innumerable "social variables" might also confound results. For example, areas with very different levels of development or infrastructure will respond differently to changes in price. These factors are sometimes included directly, sometimes handled more crudely with time trend variables.

Tests of agricultural supply laws also bring expectations into the causal

---

[11] Work up to 1977 is surveyed by Askari and Cummings (1977).

process. Price obviously should influence supply through the plans of producers. So Nerlove (1979), a pioneer in this area, tested causal processes involving not just price but also expected price. He developed a model of "adaptive expectations" making price expectations a simple function of past prices. It was these models of adaptive expectations that in part instigated the current move to "rational expectations" models that are based on *all* available information. Nerlove himself helped develop empirical models of agricultural models with rational expectations.

These increasingly sophisticated attempts to handle expectations are important for our story because they also increase the warrant of the supply laws. Expectations are a potentially confounding variable, so controlling for them adds plausibility. The supply laws have been tested extensively with the adaptive expectations variables and more recently with rational expectations variables (see Sheffrin 1983, ch. 5). The postulated causal relation between supply and price persists when either variable is included. Thus these combined tests suggest that the supply laws are not fragile – they hold across multiple specifications of their *ceteris paribus* clauses. By adding in and then varying the variable for expectations, economists are providing independent tests, fair tests, and cross tests.[12]

Thus the laws of supply seem relatively well confirmed for agricultural commodities. Although agriculture markets are ideal for such studies – the technological details are well known and extensive markets exist – there is no reason to think the laws of supply hold only for agricultural commodities.[13] There is extensive data showing that firms are generally price and cost sensitive in the ways we would expect according to the law of supply.

So economists have produced extensive data showing that the law of demand holds in consumer markets and the law of supply in agricultural markets. Those data are the result of extensive cross testing and careful efforts to show that the relevant *ceteris paribus* clauses do not hide confounding factors. This work shows that economics is no exception to the naturalism defended throughout this book.

I conclude my defense of the laws of supply and demand by considering some potential objections. Two initial concerns about my results are (1) that the work cited here runs afoul of the obstacles to belief-desire expla-

---

[12] My point here is not to defend the rational expectations literature, only to note that use of expectations variables does provide another set of background assumptions resulting in another cross test of the supply laws.

[13] Rosenberg claims that agricultural economics succeeds precisely because it relies on so much non-economic information. Yet our discussion here does not support that claim – since the extensive work on expectations, on identifying substitutes, and so on is work on economic variables.

nation noted by Rosenberg and (2) that aggregate evidence can lead to shaky inferences. We can remove both concerns by arguing a single point: the laws of supply and demand defended here are about the aggregate behavior of markets, not individual behavior. The supply-and-demand explanations defended above neither presuppose nor entail any very specific account of individual behavior (Nelson 1989). We cannot deduce from an aggregate, downward-sloping market demand curve any specific claim about how individuals relate price and quantity, because aggregate relations can result from diverse distributions of individual behavior. And even if we knew that all individuals maximized transitive preferences subject to budget constraints, we could not infer that markets as a whole would behave similarly, at least without further strong assumptions.[14] So the evidence cited in this section is evidence about large-scale behavior and that evidence warrants laws of aggregate behavior.

This means that Rosenberg's problem of the closed circle of belief and desire causes no fundamental difficulties. Belief and desire make no real appearance in the aggregate supply-and-demand laws. At most, we have aggregate variables which are rough macrolevel analogues of individual preferences and knowledge. "Expectations" are represented by the facts about past aggregate prices and quantities; what are called changes in "preferences" are really changes in the aggregate quantity consumed at a fixed price and aggregate income level. Market "demands" are really aggregate quantities purchased. These variables we can measure directly and independently, and individual preferences and beliefs have no necessary role. So even if Rosenberg's doubts were well grounded for individuals – contra the results of Chapter 6 – they would have no force here.

The aggregate nature of supply-and-demand explanations also entails that aggregate evidence is not particularly troublesome. As we saw in Section 5.2.4, the biggest worry about aggregate evidence comes when we wish to make inferences about *individuals*. However, the laws of supply and demand defended here make no such inferences.[15] Rather they are about the causal relation between aggregate demand, supply, and price. So evidence about aggregates is precisely what is needed.

Can the laws of supply and demand really be confirmed as aggregate

[14] This problem is known as the aggregation problem. See Deaton and Muellbauer (1980) for details.
[15] This is clearest for Stone's work; other tests of supply-and-demand laws that estimate the supply and demand for many commodities simultaneously sometimes do use assumptions about individual behavior to reduce the degrees of freedom of their equations to the point where estimation is possible. These efforts thus treat the market as a whole as if it were a representative individual, a presumption that is not well grounded. See Deaton and Muellbauer (1980).

laws with aggregate evidence? Don't we need their underpinning in neoclassical rationality postulates for them to be well confirmed? Our previous discussion of mechanisms and confirmation shows these questions are ill formed. If the supply and demand laws presupposed very specific mechanisms, then evidence would be called for. If we had a very well-confirmed account of economic psychology and if the evidence for the aggregate supply and demand laws was quite weak, then evidence about mechanisms would be called for. But this is not the case.

The aggregate supply and demand laws are compatible with various plausible mechanisms at the individual level. If individuals and firms are *relatively* self-interested on average, if they are *moderately* aware of price differences, if markets are at least *minimally* competitive, and so on, then we would expect changes in relative price to cause the appropriate changes in demand and supply. In short, the thoroughgoing rationality of the neoclassical postulates is unnecessary. In fact, as Gary Becker (1976b) has shown, it is so unnecessary that irrational or habitual behavior will be enough to produce the laws of supply and demand if individuals and corporations are sufficiently constrained by relative prices. Finite resources and changes in relative prices mean changes in what *can* be bought. Even a consumer who picks randomly will pick less of a good whose price has increased. So the aggregate supply-and-demand laws do not rest on very specific assumptions about individual behavior.

A final doubt about the work reported here concerns the *stability* of its results. There are several reasons to worry about stability. If the coefficients in supply-and-demand equations change over time or across samples, we might worry about falsifiability: failures of supply-and-demand laws can always be explained away by changes in the relative causal influence of other variables such as income or by changes in taste. A different objection is that the supply-and-demand laws are not real or do not identify fundamental causes if the responsiveness of supply to demand often varies.

The work of Stone and his successors cited above does not depend on ad hoc appeals to changes in taste or other such variables. Stone's work tested for changes in taste over the period of his data, but failed to find it. The very recent work of Pollack and Wales examines models explicitly incorporating factors influencing tastes. The law of demand holds up here as well. Perhaps some empirical work in these areas is vitiated by ad hoc appeals to changing parameters, but I see no evidence that all of it is or must be.

Of course, actual elasticities and other coefficients do vary for some commodities across time periods and data sets. *However,* this fact does not show that the tests involved are ad hoc, for (1) differences in underlying parameters are not invoked to explain away conflicting data and (2) the *signs* of the coefficients are stable, and that is all we need to establish the laws of supply and demand. In fact, we should be surprised to see stable coefficients across different times and markets – at least if the picture of explanation defended in this book is plausible. On that view supply-and-demand explanations identify real causes and warrant causal generalizations. But explanations of particular markets at particular times are likely to invoke unique constellations of causes; there is no reason to think aggregate demand and price have exactly the same quantitative effect in all contexts. That does not mean we cannot pick out real tendencies or partial causes, any more than variations in the fitness value of a trait across environments mean the trait is no causal factor. Stability of coefficients is not a necessary condition for successful confirmation or explanation.

7.2.2 *The central role of supply-and-demand arguments*

Skeptics about economic science may challenge the results of the last section on at least four grounds: (1) the laws of supply and demand are part of neo-classical theory and inherit its problems; (2) the laws of supply and demand are generic laws that have been known since at least Adam Smith but have gone unimproved, thus failing to meet a necessary condition for scientific adequacy; (3) the laws of supply and demand would seem to be only a small part of economics and thus my defense is of minimal significance; and (4) the laws of supply and demand, once separated from general equilibrium theories (GETs), do not explain because they are only phenomenological laws, not a theory. In this section I meet these challenges.

The supply-and-demand explanations defended here do not depend on the abstract models found in GETs or their relatives. Aggregate supply-and-demand explanations do not assume that suppliers always maximize profits, that individuals maximize utility, that prices for all commodities at all times exist, that buyers and sellers have complete information, that preferences are fully formed and ordered, or that commodities are infinitely divisible. But these are precisely the kinds of assumptions typical of GETs and of what Daniel Hausman calls "equilibrium theories." Thus the laws of supply and demand – as laws about market behavior as a whole – can be tested directly without invoking the highly unrealistic assumptions of abstract neo-classical models. In particular, the supply-and-demand explanations I defend do not depend on neo-classical consumption theory, a theory that has serious empirical failings as a theory of market behavior.

What then is the relation between the basic laws of markets and neo-classical theory? I suggest the following picture. The laws of supply and demand are close to what are sometimes called *phenomenological laws,* a kind of law which is usually contrasted with *theoretical laws.* Phenomenological laws describe – either in causal or non-causal terms – the aggregate behavior of relatively observable quantities; theoretical laws provide a more abstract characterization that explains why those phenomenological laws hold, frequently by citing underlying unobservable mechanisms. The gas laws relating pressure, temperature, and volume are phenomenological laws; statistical mechanics provides the corresponding theoretical laws. The laws of supply and demand are like phenomenological laws because they describe the aggregate behavior of markets. Like the other social science "laws" we defended earlier, they describe a general causal process and identify some, but not all, of the complicating variables, variables which get filled in case by case. On the other hand, the models of GET or equilibrium theories in economics postulate such relatively unobservable entities as preference orderings, production sets, and so on. From these basic postulates they try to derive aggregate market behavior. In short, the models of GETs describe theoretical laws about the alleged mechanism behind aggregate market behavior. GETs would, if well confirmed, relate to the laws of supply and demand much as statistical mechanics relates to the gas laws.

Viewed this way, it makes good sense that GETs and supply-and-demand analysis should be largely independent. GETs try to provide theoretical mechanisms. But as we saw in Chapter 5, we can confirm causal laws without supplying mechanisms. Mechanisms are important if our understanding of the mechanisms is better than our understanding of the process they bring about or if our macrolevel theory presupposes some specific mechanisms. But, as we saw above, this is not the case for aggregate supply-and-demand explanations. So the laws of supply and demand do not inherit the problems and puzzles associated with GETs.

Let's turn next to the claim that supply-and-demand explanations involve generic laws that have been around since Smith. We can answer these doubts on several fronts: (1) increasing predictive success is not a necessary condition for scientific adequacy, (2) generic predictions are no sign of poor science nor are they unimprovable, and (3) the laws of supply and demand do show a history of increased predictive success. Increasing predictive success is surely a scientific virtue, though its precise meaning is not at all obvious. Nonetheless, increasing predictive success seems neither necessary nor sufficient for good science. Take a stupid theory that is far from the truth as compared to a theory that starts out, as it were, with basically the right account of reality. Which theory is likely to show increasing predictive success? Any improvement in the stupid theory will drastically increase its predictive success; the basically correct theory, however, will find improvement hard to come by. Thus increasing predictive accuracy is just one component in theory assessment; more static facts about the current state of the evidence can override it.

Furthermore, generic predictions *can* make for good science. As I argued in Chapter 3 and elsewhere, much biology identifies basic causal processes without specifying in detail all the relevant variables. These causal processes are identified *in general;* we are not given their precise strengths. Evolutionary biology, for example, cites the main causal processes, but it has to work out case by case how those processes combine. Even then, predictions are still frequently about the direction of causes, not their precise influence. Grant's exemplary work, which we discussed in Chapter 3, is full of "generic" predictions. Moreover, molecular biology – a hard science if there ever was one – likewise relies heavily on generic predictions. It tells us that signal sequences cause proteins to sort to different parts of the cell, that second messengers convey external signals to the nucleus, that receptors recognize external signals, and so on. These causal claims are generic – they describe causes but not quantitatively. They do not tell us how quickly proteins are sorted, how the rate of sorting is influenced by the rate of protein syntheses, and so on. Quantitative formulas are rare indeed in cell biology.[16] So unless we are willing to relegate our best biology to the category of failed or inferior sciences, generic predictions are no essential sign of poor science.

Equally important, generic predictions can be improved. We can do so, for example, by identifying further causal factors, by differentiating the contexts in which factors work, by filling in the factors responsible for exceptions, and so on – all in non-quantitative terms. Providing these more detailed but generic predictions will presumably give us better predictive power. And, as I argued in Chapter 5, we can improve generic predictions in this way while staying at the aggregate level, as happened with the laws of supply and demand as further aggregate variables were incorporated.

Finally, economists have refined the laws of supply and demand. It is true that Smith identified supply-and-demand processes.[17] Yet Darwin also described the basic mechanism of natural selection, but we hardly conclude that therefore there has been no progress in evolutionary biology. In both cases, the initial processes were described in verbal form, with some informal and generally anecdotal evidence and with some suggestive remarks about how to apply them to more troubling cases. As with the theory of natural selection, modern economics has provided far more extensive and sophisticated quantitative data than the founding fathers of the discipline could have imagined.

---

[16] As a recent *Nature* editorial points out (Maddox 1992).
[17] Though it is doubtful he conceived them as they have been developed in the twentieth century (Mirowski, personal communication).

The history of research on consumer demand cited earlier illustrates this point nicely: (1) Smith's anecdotal evidence was replaced by quantitative evidence of much higher quality in the early part of this century; (2) Stone and others significantly refined that work both by more carefully sorting out complicating variables such as the income effect and by providing still more sophisticated statistical evidence; (3) numerous investigators replicated Stone's work with different data sets under different statistical assumptions; and (4) recent work has added in further complications associated with demographic variables, interdependent preferences, and departures from equilibrium, all the while showing that the law of demand holds up. Each of these steps provides predictive improvement, the kind typical of sciences that begin by describing causal processes in rough outline and then move to more detailed specifications. As I have been arguing throughout this book, much good natural science proceeds in just this way. Perhaps the abstract models of neo-classical economics have not shown a similar history of predictive success and that Rosenberg's concerns are thus reasonable in that domain, but those doubts are not compelling when it comes to supply-and-demand explanations.

That said, supply-and-demand work sometimes *does* provide precise quantitative claims about relative influence. Innumerable studies have been done on particular commodities in particular markets over particular periods. In these defined contexts, much more quantitative predictions are possible as the full details about complements, substitutes, and other generic factors are filled in. As Rosenberg grants, the work cited earlier in agricultural economics is a paradigm example. So generic predictions are no fundamental obstacle to the work I have defended here.

Let's turn now to a more serious doubt – namely, that the laws of supply and demand are only a small part of economic explanation and thus that my arguments have limited significance even if successful. It is true that much modern economics revolves around general equilibrium theory. But I also believe that the basic supply-and-demand laws provide an explanatory scheme that undergirds much useful work in economics. Commentators have focused on GET for a variety of reasons: because it is what gets taught in theory classes, because it fits our leftover positivist assumptions that good science is always embodied in formalized theories and that theories capture most of what is important in science, and so on. However, equating modern economics with GET and the like ignores important parts of the discipline – frequently those parts where supply-and-demand explanations are most important.[18]

---

[18] It is here that I disagree with Hausman's approach, despite my general debts to his work on economics.

We can see the important role that supply-and-demand analysis – sans GET or equilibrium theory – plays by continuing our analogy with evolutionary theory. Philip Kitcher (1985) has argued that the core of evolutionary theory is the "Darwinian history" – the explanation of a biological trait by identifying the relevant ancestor species and environmental factors and then applying the mechanism of natural selection to show how that trait arose. Darwinian histories involve applying the same basic argument pattern or schema over and over again, filling in the schema with the specific details of each case. The basic strategy always involves three rough steps: (a) identifying the relevant trait and (b) applying the mechanism of natural selection by (c) identifying the basic factors underlying the selective environment and the fitness of traits. It is this argument pattern, rather than a simple set of propositions or an abstract theoretical model, that unifies Darwinian theory and directs its research.

I would argue that the laws of supply and demand do something similar for economics: they provide a general schematic strategy that is applied again and again to diverse phenomena by fleshing out contextual details. That general schema seems to me to involve something like the following steps:

(1) Identify the relevant commodity, market (buyers and sellers), and measure of price.
(2) Explain that market by determining the rough slope of the demand and supply curves and by identifying both changes along and shifts in the demand and supply curves as described by the laws of supply and demand.
(3) Identify the basic factors determining shifts in supply and demand curves by looking at changes in the assets and costs of buyers and sellers, particularly as influenced by the price of substitutes (equivalent goods to the one under study) and complements (other goods required by the production or consumption of the good under study).

The first step often calls for real creativity, perhaps because the commodity has to be defined, or because there is no public price and a shadow price has to be identified to determine exchange rates, and so on. This first step is roughly analogous to identifying the relevant trait in a Darwinian history. The second step is applying the basic mechanism, which requires fleshing out the forces involved according to context in the third step. Like the Darwinian history, the supply-and-demand strategy only specifies in rough terms the basic kinds of factors that underlie supply-and-demand forces. It is again a creative process to pick out the relevant substitutes and complements, assets, and costs.

This basic argument pattern turns up again and again as economists seek to explain new phenomena or more fully analyze standard ones. Let me mention some obvious applications. The supply-and-demand argument pattern is applied both "vertically" and "horizontally" – both to smaller and smaller subsections of a more inclusive market and to new phenomena not previously covered. Some typical examples include the markets for:[19]

Horizontal Applications	*Vertical Applications*
medical interns	wilderness areas
credit cards	"lemons"
managerial reputations	illegal goods
exports and imports	health
swap goods	information
overdrafts	education

Of course these supply-and-demand accounts vary in empirical success. However, my concern now is not empirical adequacy – something I argued for earlier – but showing that the supply-and-demand argument strategy plays an important role in economic explanation. The examples listed above and innumerable others like them strongly support that thesis.

I want to turn finally to doubts that the laws of supply and demand constitute an explanatory *theory*. These laws are, on my view, phenomenological laws. They cite general causal processes or tendencies of aggregate market behavior; when combined with local detail, they provide reasonably full accounts of the large-scale causal processes. Nonetheless, supply-and-demand explanations produce only low-level aggregate generalizations, not a full-fledged mathematicized theory replete with laws of underlying mechanisms. And it is precisely the latter which GETs or equilibrium theories deem to provide. By separating supply-and-demand processes from abstract economic models, haven't I defended the least explanatory part of the discipline?

The immediate discussion above and the arguments from Chapter 3 on theories and explanation show why this objection fails. Phenomenological laws can explain. They can cite causes, and when they are well confirmed,

---

[19] Some typical instances include B. Brown (1989) on swap markets; Yang (1987) on exports and imports; Pavel and Binkey (1987) on credit cards; A. Roth (1991) on interns; Becker, Grossman, and Murphy on drugs (1991); and Akerloff (1970) on "lemons."

they show that the factors cited are at least partial causes or tendencies. So phenomenological laws can explain, regardless of whether they are embedded in a larger formalized theory – just as the gas laws explained before they were embedded in statistical mechanics. And as we saw in Chapter 5, causal generalizations can explain without identifying underlying mechanisms. As I have argued repeatedly, much explanation in the biological science follows just this pattern. Furthermore, suppose that the causal view is wrong and that unification is the key to explanation, despite the arguments of Chapter 3. The laws of supply and demand will still fare well, for reasons described above: they provide an argument strategy or pattern of widespread application. On the most developed account of explanation as unification – Kitcher's – that should make them explanatory as well. So defending the laws of supply and demand *is* defending a significant part of economic explanation.

7.3 Assessing neo-classical models

My discussion so far has left an apparently important question unanswered: What is the status of standard neo-classical theory? Much work in philosophy of economics has tried to answer that question. In this section I want to make some brief remarks about this debate.

If the arguments of the previous sections are right, then to some extent the debate over neo-classical theory is misplaced. Supply-and-demand explanations can proceed without neo-classical theory. Of course, the neo-classical paradigm does play an important role in the sociology of the profession; it no doubt plays important heuristic roles as well. Still, when we look at the *evidence* for supply-and-demand explanations and the way economists flesh out those explanations, GETs and the like are of minor importance. Since supply-and-demand explanations are the bread and butter of applied economic analysis, this means that GET has a secondary role for a large body of economic science. To that extent focusing the debate on the status of neo-classical theory is misplaced.

Asking about the status of neo-classical theory is misplaced in another way. Pursuing this question assumes there is "the" neo-classical theory and "the" evidence for it. Both assumptions are unhelpful. As we pointed out in Chapter 2, a "theory" is frequently a class of related but independent models of the world. Abstract models in economics appear in many different contexts and for many different purposes. "Neo-classical theory" has many different guises in different contexts, and it seems a leftover positivist assumption to think that these diverse practices are capturable in one single formulation. Hausman's (1992) effort to distinguish equilibrium theory from its augmentation into general equilibrium theory takes a step towards a more realistic picture of complex economic practice, but it does not go far enough. Hausman's equilibrium theory, which is supposed to be the "real" core of neo-classical economics, also imposes more structure than is justified. Much exciting work in the neo-classical tradition rejects profit maximization by the firm as a whole for a more realistic and more individualistic account that treats firm behavior as the outcome of individuals pursuing their self-interests, an approach that in no way assures profit maximizing for the corporation as a whole.[20] We also saw in previous sections that disequilibrium models are at the cutting edge of current research. So two key components of Hausman's "equilibrium theory" – profit maximization and equilibrium – are conspicuously missing from current neo-classical practice. This suggests that trying to identify "the fundamental theory" is misguided.

Similar conclusions should hold for questions about the evidential status of neo-classical theory. The postulates in Hausman's version of equilibrium theory are unlikely to be either true or false, well confirmed or disconfirmed. Instead they will be relatively reasonable in some contexts and indeterminate or implausible in others. For example, when corporate ownership and control are not separate and when the level of competition is high, profit maximization may be a reasonable assumption – and at the same time may be a serious distortion when these factors are not present. So asking about *the* cognitive status of the theory is the wrong way to frame the question.

Thus I am unpersuaded by Hausman's general defense of equilibrium theory. Though he thinks that economists are dogmatic in their belief that economics must be a separate science, he thinks that holding onto equilibrium theory in the face of empirical failure is reasonable, given the uncertainties of econometric testing. I doubt that any such overall evaluation will be reasonable. Econometric evidence varies in quality from case to case as do the substantive auxiliary assumptions that must be made to test equilibrium theory. We can reasonably ignore some failures; others – especially those that are replicated across different statistical specifications – we cannot. Moreover, economists often use equilibrium theories to explain and make policy recommendations while having very little evidence that their models make reasonable assumptions.[21] So to evaluate the neo-classical tradition, we must proceed case by case, looking at the evidence, the *ceteris paribus* clauses, the substantive auxiliary assumptions needed to tie the theory to reality, and so on (an approach Hausman him-

---

[20] For a discussion of the strengths and weaknesses of some of this work, see Kincaid (1995).

[21] I discuss one instance of this practice in work on the theory of the firm in Kincaid (1995).

self recommends in a different context and in fact practices with great success). Only after such studies dare we generalize about the status of neo-classical theory, and even then uniform assessments are unlikely. My own sense is that far too much of neo-classical economics is dominated by abstract model building, work done at the expense of carefully developing and testing realistic, explanatory models. But backing up this assertion requires much more detailed analysis than I have done here.

Evaluating abstract models in economics requires a longer, more nuanced story for another reason. Science involves more than theories and evidence – it involves problem-solving strategies, norms, methodologies, social processes, and more. If neo-classical models fail to explain, they might nonetheless serve other important functions. Unrealistic models are endemic in good science, and many of them are never intended to be explanatory by themselves. Rather, those models do other things. They are sometimes useful for predicting, though not for explaining. They often play important heuristic roles. They can help suggest when a result might not be sturdy, by seeing how changing a model's assumptions affects its performance. By helping us to draw out implications, abstract models can also suggest new hypotheses or possible relations in the data worth exploring. They can make complex phenomena mathematically tractable and thus set the stage for ultimately developing realistic and explanatory theories. Abstract models can provide a useful tool for suggesting realistic theories and their complicating variables. These are just some of the more obvious functions that models play – observers of science have much to learn about how abstract models function in good science.[22]

Do the abstract models of neo-classical economics successfully serve these non-explanatory functions? Again, no easy answer is possible – we need to look at specific models in specific contexts. When equilibrium models serve to organize, generate, and suggest important variables for supply-and-demand analysis, then they inherit at least that instrumental success. Furthermore, the constrained maximization approach has time and again suggested potential models for previously unstudied economic phenomena. However, if abstract economic models persist because they serve other functions – such as fitting with the reward structure of the discipline or the political ideology of economists – then they are indefensible. A full evaluation of neo-classical models thus requires a detailed and nuanced account, something I cannot pretend to give here.

So I raise the question of neo-classical economics only to avoid answering it. Weak defenses and unfair attacks can perhaps be defused, but any

---

[22] Biologists and philosophers of biology have been among the few people to think about these issues in concrete terms. See M. Becker (1959) and Levins (1966).

more compelling assessment requires a much more complex discussion than I can give here. At best I have offered a strategy for pursuing the question.

7.4 Reduction and microfoundations

Economists espouse methodological individualism. Numerous commentators have taken them at their word and argued that our best social science – modern economics – thus supports the individualist program. Even substantive debates in macroeconomics are claimed to turn in part on such issues: defenders of rational expectations, for example, claim that no macroeconomic process is understood until we have the microfoundations. In this section I use the tools of Chapter 5 to look at these issues.

The most common individualist theses in economics are the following:[23]

(1) All economic theory is reducible to a theory about individuals.
(2) All macroeconomic results are derivable from microeconomic theory.
(3) Any well-confirmed and explanatory macroeconomic theory must be consistent with microeconomic theory.
(4) Any well-confirmed and explanatory macroeconomic theory must have a complete foundation in microeconomic theory.

The first claim is the standard reductionist thesis we discussed in detail in Chapter 5. Economists asserting the second thesis apparently claim both that the microeconomic facts determine the macroeconomic facts and that macroeconomic processes are fully explained once we explain the individual behavior realizing them. So thesis (2) is a version of the individualist claim that lower-level theories suffice to fully explain regardless of reducibility (a thesis discussed in Section 5.2.1). Theses (3) and (4) are about mechanisms. So the general arguments of Chapter 5 will thus be directly relevant.

Let's look first at the idea that economics is in principle reducible to a theory about individuals. Not surprisingly, I think this thesis implausible. Doubts immediately surface when we ask what is supposed to be the *reducing* theory. The obvious answer is microeconomics. Yet that answer will not do. Microeconomics is not a theory about individuals. This point is straightforward, yet it is frequently missed.[24] The basic entities of microeconomics are in most formulations households and corporations. But

---

[23] The best single source for documenting these claims is Weintraub (1979).
[24] For example, Boland (1982, p. 80).

households and corporations are social entities. They may be "smaller" than other social variables, but they are surely social.[25] So even if – and we shall have reason to question this "if" later – macroeconomics was reducible to microeconomics, individualism would not win the day.

Perhaps microeconomic reference to "households" and "corporations" is only an eliminable *façon de parler?* This is unlikely. There is no extant economic theory that comes even close to reducing talk about corporations and households to that about individuals. Economists have recently begun looking at the dynamics underlying household choice and corporate behavior.[26] But that work is in its infancy, and we have reason to doubt that it will produce *reductions*. One obvious problem is that microeconomic predicates or kinds are prone to multiple realizations in individual terms. Assume for the moment that corporations do maximize profit as traditional microeconomic theory claims. Profit maximizing can happen via numerous different organizational structures and individual behavior. "Profit maximizing" describes a kind of behavior; it abstracts from the individual detail. So no manageable set of individual behaviors is likely to capture all and only the corporate search for profits. Humans are creative, and economic selection does not care, as it were, how corporations maximize – it only (allegedly) ensures that they do.

If corporations pursue more than profits, then reduction is even more unlikely, for we now have multiple charactertistics to reduce. Moreover, a standard defense of the profit maximizing postulate – the economic selection argument – itself raises further problems. Nelson and Winter's (1982) work, which is the most elaborate effort to develop a realistic model of economic selection, argues that if firms are profit seekers, they may be so only as a result of pursuing other corporate goals. Thus profit seeking itself may be a behavioral description that collates diverse corporate strategies, each of which no doubt comes about by diverse individual behavior. For example, profit maximizing behavior can result from hierarchical and non-hierarchical structures, from owner-managers maximizing their self-interest, from managers who are not owners but disciplined by the market for managerial talent, by managers who are not owners but who pursue profit because it maximizes their internal political power, and so on. So multiple realizations occur at multiple levels. Finally, any theory of individual behavior in corporations is likely to identify individuals by their

---

[25] Some (Schrader 1993, Janssen 1993) allow individualist theories to refer to social institutions, etc., if they treat them as fundamental entities; this would transform Hegel and other notorious holists into methodological individualists and seems implausible for that reason.

[26] Work on the economic processes inside firms is now a booming industry. For a survey of recent work, see Williamson and Winter (1991).

corporate role. Thus the problem of presupposing social information is probable as well.[27] (It should be clear how similar arguments could be made about households, so I will not pursue these problems further.) For all these reasons, individualist theories are unlikely to capture important patterns at the level of the firm.

This result also makes reducing *macroeconomics* quite unlikely as well. Any reduction of macroeconomics is likely to be a two-stage process. We first reduce macroeconomics to microeconomics and then reduce microeconomics to a theory about individuals. But we have just shown that this second step is highly unlikely. That said, a more restricted reductionist thesis might be plausible – namely, that macroeconomic explanations are reducible to our best theory of corporate and household behavior.

One immediate obstacle to this more limited thesis is that government behavior is an important macroeconomic variable. So a full reduction would demand that we eliminate reference to such behavior in favor of talk about individuals. Economists have had insightful things to say about how self-interested individual behavior influences government activity. Yet that work falls far short of being a full reduction, for reasons familiar from our discussion of rational choice theory in Chapter 5. So any plausible reductionist claim about macroeconomics must be even more limited – it must claim only that the *purely economic* parts of macroeconomic explanations are reducible.

To make any firm judgment about this more limited reductionist thesis, we would need to know what parts of macroeconomics, if any, are currently well confirmed – and make an argument that they and their likely successors are likely to be irreducible. This chapter, however, is only an outline of how naturalism and holism can be applied to economics, and I cannot pretend here to make even rough judgments about what parts of current macroeconomics succeed and what do not.[28] So any conclusions will be tentative.

That said, we can outline some reasons why good macroeconomics might well be irreducible to a good microeconomic counterpart. Those reasons are the general obstacles to reduction cited above and in Chapter

---

[27] Other reasons can be constructed. Imagine, as seems reasonable, that household and corporate behavior result from complex bargaining between individuals. We know from game theory that most complex games admit of multiple outcomes and that similar outcomes can result from different reward structures and bargaining paths.

[28] Though a general approach is suggested by earlier results in this book – namely, that macroeconomics produces some relatively well-confirmed causal generalizations, generalizations that are generated by applying the basic supply-and-demand argument strategy at a very aggregative level.

5, in particular the problems of multiple realizations and presupposing information. Both may well make reducing macroeconomics to microeconomics difficult.

Macroeconomics by nature studies economic behavior at a high level of aggregation. It thus looks for general patterns among large-scale entities, abstracting from the many specific low-level causal particulars. That means multiple realizations are at least a possibility, for large-scale patterns that hold across diverse times, economies, and so on might not be the result of a common causal process at the level of firms and households. While macroeconomics describes markets at their greatest aggregation, those markets result from many "smaller markets" of various kinds. For example, the money market is composed of the markets for bonds, credit cards, cash balances, and so on. And those markets are in turn brought about by different combinations of corporate and consumer behavior. As a result, there seems plenty of room for the same macroeconomic patterns to be brought about by diverse underlying processes – either in the smaller markets realizing them or by corporate and household behavior. A shift in supply schedules in the money market, for example, could potentially result from various different changes in submarkets and those in turn from diverse kinds of corporate and household behavior. So multiple realizations may well be a real possibility.

We can also make a tentative case that explanations in terms of corporate and household behavior are likely to presuppose macroeconomic theory rather than reduce it. Macroeconomic variables play various roles inside microeconomics itself. The price of inputs influences the supply curve of producers. Explaining those prices, however, may sometimes require citing macroeconomic variables and processes. The price of loans, for example, will depend on the interest rate, a factor determined in the large-scale money market. Wages will be set in part by reference to the consumer price index, another macroeconomic variable. In fact, attempts to incorporate explicit expectations in microeconomic accounts would seem to give macroeconomic variables a rather essential place. Insofar as economic agents plan in expectation of large-scale macroeconomic states – the growth rate, the prospects for recession, and so on – those variables are part of microeconomic explanation; they causally influence microeconomic processes through the beliefs of individuals.[29]

---

[29] As usual, individualists could hope to eliminate these whole-to-part causes in favor of causes relating only to individuals. But, as we saw in Chapter 5, such reducibility must (1) be argued for and (2) avoid all the standard obstacles to reduction. In short, appeal here to what could be done is simply a pious hope in face of the actual obstacles to reduction outlined above.

Finally, macroeconomic variables may be involved in determining the values of microeconomic variables, as Alan Nelson (1984) has argued. We saw in discussing the law of demand that income, for example, is a relevant causal factor. However, most data on income are aggregate data, involving total income changes, not changes in household income household by household. This means that either we explicitly incorporate a macroeconomic variable in our microeconomic account or we use the macroeconomic variable to infer the microeconomic one. In the former case, we obviously have not eliminated but presupposed at least some parts of macroeconomic theory. In the latter case, it may appear that we are using macroeconomic variables only as a tool, not for explanatory purposes. Nelson, however, argues that if macroeconomic variables are reliable indicators of microeconomic ones, then there must be a causal relationship involving macro- and microeconomic variables – one that is being assumed, not reduced. In either case, macroeconomic theory is not reduced by microeconomic accounts.

So we have a prima facie case against reducibility in economics – with the strongest case probably being against reducing all economics to individual (as opposed to firm and household) behavior. We need now to turn to other individualist theses in economics, ones economists frequently do not clearly separate from strict reduction.

Economists sometimes claim that any macroeconomic result can be gotten from microeconomics. In other words, microeconomics ideally should *suffice* to explain all economic phenomena. There are several reasons for thinking this unlikely. For one, those who assert this claim often believe that current microeconomics in the form of GETs or equilibrium theories can play such a role. Putting questions about confirmation aside, these theories are incomplete – they give little or no account of dynamic processes. But macroeconomics is also about processes, not static equilibrium analysis. So in their current form, microeconomic theories are perhaps unlikely to capture all important macroeconomic phenomena for this reason alone.

Would the ideal, fully developed microeconomics suffice to explain all macroeconomic phenomena? We gave several arguments in Chapter 5 that argue in the negative. Lower-level theories are likely to presuppose higher-level information. When that happens, there is no prospect for them to completely explain higher-level processes. However, we saw earlier in this chapter that microeconomic accounts might well presuppose at least some explanation of macroeconomic variables. If so, microeconomics could not fully explain. Furthermore, even if each particular macroeconomic event had an explanation in purely microeconomic terms, that would not mean all macroeconomic explanations can be captured. If macroeconomics is irreducible as I argued above, then it will capture broad causal connections between kinds of events not visible from the microeconomic perspective. In short, while microeconomics might explain case by case, there would still remain an essential role for macroeconomics to explain basic *kinds* of economic processes at the aggregate level – assuming that my arguments against strict reducibility succeed. Since I spelled out these general arguments in careful detail in Section 5.2.1, I will not develop them further here.

Let's turn next to the common idea that macroeconomics must have microfoundations. At issue is whether macroeconomic claims could be relatively well confirmed without spelling out the microeconomic mechanisms. To answer this question, we can apply the results of Chapter 5, especially Section 5.2.3. We argued there that there is no *general* requirement for lower-level mechanisms. Instead, in each context we must evaluate the relative plausibility of our lower- and higher-level theories. If we have a very well-confirmed theory at the lower level, then the demand for mechanism carries weight. If our higher-level theory is well confirmed and our lower-level theory is not, then the situation is reversed.

Frequently those demanding microfoundations want a foundation in GET or at least rational maximizing behavior. I doubt that such demands are warranted – because GET models and their variants are not particularly well confirmed. Thus they do not make for very strong constraints on macroeconomic theorizing. In fact, some parts of macroeconomics may have more empirical merit than abstract equilibrium models, and thus the real constraint perhaps should run from macro- to microeconomic theorizing. Of course, given my defense of supply-and-demand accounts, I am considerably more sympathetic to another kind of microfoundations – namely, showing how large-scale processes come about via less aggregate supply-and-demand forces.

Thus when defenders of rational expectations criticize macroeconomics for lack of microfoundations, there is both truth and falsity in their complaints. Rational expectations as an account of individual behavior is not a very well-confirmed theory, probably not as well confirmed as other parts of microeconomics. And to demand rational expectational mechanisms on the ground that the rational maximizing model is central to economics is, on the perspective defended here, to confuse the role of a heuristic model with an explanatory theory. Moreover, it certainly does not follow – as some (Begg 1982, p. 5) in the rational expectations literature claim – that every macroeconomic theory must include an expectations explanation because expectations are part of the causal process. After all, the genetic endowment of individuals is also ultimately part of the macroeconomic process, but surely we need not include it to have well-confirmed, explanatory macroeconomic accounts. Again, how much detail we need about microlevel processes varies according to context – according to what our macrolevel theory presupposes, how well confirmed it is, how well confirmed our microeconomic account is, and so on.

There is nonetheless truth in the call for microeconomic mechanisms. If a macroeconomic theory makes strong assumptions about individual behavior, then we may need evidence for those assumptions before the theory is empirically adequate. If a macroeconomic theory *essentially* presupposes that individuals consistently and obviously are fooled when real chances for learning occur, that account is implausible. Yet, avoiding this implausible presumption is not equivalent to demanding the extreme rationality postulates associated with rational expectations. For example, individuals can avoid repeated stupid mistakes without having, or behaving as if they have, the full or best possible theory of all economic variables, a typical assumption of rational expectations models in macroeconomics.

Similar points hold for the laws of supply and demand taken as laws about aggregates. Those laws must be consistent with a *plausible* account of individual economic behavior, to the extent that we have such an account. Yet that constraint is a weak one. The laws of supply and demand, interpreted as laws of aggregate or large-scale market behavior, are consistent with a quite wide range of individual-level processes. As I pointed out earlier, Becker has shown that even random choice by individuals will produce a downward-sloping demand curve if those choices face a budget constraint. Thus identifying individualist mechanisms for the laws of supply and demand is no pressing requirement for confirmation. Of course, individualist mechanisms would be welcome and helpful, but that is a claim few antireductionists would deny.

Are microfoundations perhaps a necessary requirement for predictive improvement, if not confirmation in general? We can again apply our results from Chapter 5. There we argued that macrosociological accounts can in principle be improved by specifying further macrosociological variables, a conclusion supported by the examples of Paige and Hannan and Freeman and the clear evidence that the natural sciences often achieve predictive improvements without invoking microlevel variables. Microfoundations, we argued, may be useful routes to predictive improvement, but there are also potential obstacles. When different individual-level behaviors bring about macrosociological processes, individualist mechanisms may not provide predictive improvement. Moreover, even when mechanisms are useful, nothing requires that they be individualist in nature – often the most obvious level of disaggregation is to further social entities.

These general points hold for economics. We saw earlier in this chapter that the laws of supply and demand have been improved over time by identifying further aggregate variables like the aggregate income effect or social variables like the level of development. Improvement can and has come without providing individualist mechanisms. No doubt there is a place for microfoundations in improving our predictions about supply and demand. The aggregate income effect, for example, can be more precisely understood by factoring in how income is distributed. Yet this role for microfoundations again leaves us with an anemic form of individualism for several obvious reasons. Since the supply-and-demand laws can be improved at their own level, mechanisms are no necessary requirement for improvement. Holists can grant the uninteresting claim – which really follows from the truism "the more evidence, the better" – that microfoundations will help confirm. Ultimately, it is an empirical question how much specifying microfoundations will increase our predictive power, though some recent studies on disaggregated demand data suggest that ignoring underlying detail is no serious loss (Blundell, Pashardes, and Weber 1993). Even if microfoundations were a necessary requirement for confirming the supply-and-demand laws, that would not support individualism. When economists provide non-aggregate evidence for the supply-and-demand laws, they typically do *not* give individualist foundations. Instead they look at the behavior of social entities such as households and firms. Thus individualist strictures in any strong form are doubly misguided.

Similar conclusions may hold for predictive improvement in macroeconomics. The natural units for disaggregating macroeconomics would be households and corporations rather than individuals. And since macroeconomic and microeconomic variables seem to be realized in diverse individual behaviors, it is unclear just how much predictive improvement of those phenomena will be gained by looking at individual detail. If corporate behavior, for example, is shaped by a selective process that does not "see" individual-level detail, emphasis on the latter may not lead to predictive improvement. Thus making individualist mechanisms a necessary requirement for predictive improvement seems misguided. Of course, substantiating such claims would require a detailed look at empirical work in macroeconomics, a task far beyond the limited goals of this chapter.

We can thus tentatively conclude that the main individualist theses gain little support from economics. My arguments have, of course, been brief, and a decisive account awaits a much fuller development of the main arguments outlined here. However, the above discussion should both make it clear how such an argument would go and give some prima facie plausibility to the claim that it could be carried out.

8

Problems and prospects

In this last chapter I conclude my argument by taking up two closely related questions: Why haven't the social sciences fared better and what can be done to improve them? In past chapters I often criticized arguments alleging that no science of society is possible. Yet many of the arguments I criticized were also put forth as *explanations* – explanations of why the social sciences have been so unsuccessful. Thus a complete defense of naturalism should also explain why the social sciences have not fared better. Identifying the causes of bad social science may also tell us how the social sciences might be improved. So this chapter is devoted to these two final questions – to identifying the obstacles to progress in the social sciences and to proposing ways to deal with them. These are big topics. Nonetheless my discussion will be brief for reasons I outline below, chief among them being that philosophical argumentation is unlikely to decide what are really complex empirical issues.

Standard philosophical explanations of poor social science rest on several false presuppositions. Most obviously, philosophers like Rosenberg assume that social science has been a dismal failure. Yet in previous chapters I have repeatedly argued that some social research – the work of Paige, of Hannan and Freeman, of Brown and Harris, and of economists on aggregate supply and demand – is relatively well confirmed. This work, I have argued, compares favorably to the best work in evolutionary biology. So while we should ask why the social sciences are not better than they are, I deny that they fail *tout court*.

Putting the question this way – as "why does *some* social research do so poorly?" – leads us down a very different path from that pursued by typical philosophical arguments on these issues. For once we deny that all social science is a failure, then we have much less reason to think there is a single universal explanation for bad social science. In particular, we have no reason to look for one, basically conceptual fact about the social sciences that explains why they fail. For example, philosophers often claim that the social sciences have failed because they have not hit upon natural kinds, unlike the natural sciences – in other words, the natural sciences describe reality in a way that cuts nature at its joints and the social sciences do not. Some such argument was behind Searle's antinaturalist views discussed in Chapter 3, behind Rosenberg's attack on belief-desire psychology, and behind many other philosophical diagnoses of social science failure.[1] Yet, if some social research actually succeeds, then the explanation for poor social research is unlikely to be that the social sciences have not found the right vocabulary.

We have other reasons to be wary of unitary, philosophical explanations. For one, these explanations themselves are weak by scientific standards. As has been shown in the debate over scientific realism, explaining the success of science by appeal to its truth – by its having cut nature at the joints – faces enormous problems. The main problems include conflicting evidence and plausible alternative explanations for the success of science. Conflicting evidence comes from the history of science, where we find relatively successful theories that we now judge to be false. So truth is not the only possible cause of success, and thus it may not be the cause of success now. Moreover, there are competing explanations for scientific success, and we do not have to tell skeptical stories about brains in a vat to construct them. Natural science is a selective process, allowing theories that meet basic scientific standards to go forward and weeding out those that do not. Furthermore, this process clearly works by selecting for empirical success, not for truth. So to conclude that natural science succeeds because it has hit upon natural kinds – and social science fails because it has not – is to provide a very weakly supported explanation indeed. No similar argument inside the natural sciences would be acceptable. It would be considering a theory as well confirmed despite the facts that (1) it was known that the data could be produced by processes other than those postulated and (2) alternative reasonable explanations of the data exist. This would be bad science, and its philosophical equivalent would be no better.

We have other reasons to doubt general, philosophical explanations. Everything we have seen in earlier chapters suggests there should be many different piecemeal explanations rather than one overarching cause. I say this in part simply because earlier chapters have seen different obstacles to good social science in different disciplines. Economics, for example,

---

[1] For example, Nelson (1990). I discuss the problems with these sorts of arguments in more detail in Kincaid (in press b).

runs into problems in the way it treats models. Anthropology, on the other hand, seldom produces elaborate models but does provide functional explanations. Those explanations, we saw, have their own unique difficulties. Much the same holds for other obstacles we pointed to (and will discuss more fully below). Moreover, if the social sciences fail in part for social reasons – because they do not have the right social structures – then we should expect the full explanation ultimately to invoke concrete detail that varies from context to context. There may be useful generalizations, but if the general picture of social explanation from previous chapters holds true, the real explanatory work will come by filling in local detail.

So how should we go about explaining poor work in the social sciences? Ideally we would proceed as follows: Following the perspective outlined in Chapter 2, we would begin from a detailed and well-confirmed philosophy of science that described the traits of good scientific products and practices and a well-confirmed sociology of science that told us what social practices produce good science. We would then use those accounts to investigate the social sciences. Looking at social scientific theories, evidence, testing, heuristics, and methods – what we might call the "cognitive level" – we would identify cognitive reasons why social science fails. Looking at the "social level," we would identify specific social practices that produced poor science. From these two results recommendations would flow.

Obviously this ideal is out of reach, for we do not have the requisite theories. So our best alternative is to suggest hypotheses whose final evaluation must await future investigation. Yet we can at least identify some relatively plausible conjectures, relying on what we know now and on the specific discussions of social science practice in previous chapters.

Let me begin with the cognitive side of science – with the epistemic causes of poor social science and the corresponding epistemic routes to progress. Perhaps two factors are crucial: the relative dearth of controlled experiments and the lack of relevant theories from outside domains to serve as background knowledge. The first point is obvious and often noted. Controlled experiments are not the only way to produce good science, as I argued in Chapter 3, nor are they impossible or unknown in the social sciences. Nonetheless their potential role is limited, and that means the most powerful route to cross tests, fair tests, and independent tests is often unavailable. Note that the difficulty of controlled experiments also haunts important parts of biology, for example, and for just that reason progress is slower there than, say, in molecular biology.

Molecular biology has shown such tremendous success not just because it has controlled experiments. What makes those experiments so powerful is that they can call on a wealth of background knowledge from outside

Problems and prospects 261

biology, primarily from chemistry. That background knowledge makes for much more powerful tests. The social sciences, however, have no such body of knowledge to call upon, and as a result, inference from data to hypotheses requires much greater leaps and results in greater uncertainty (see Lieberson 1992). Again this problem is not unique to the social sciences. Ecology and evolutionary biology have less established background knowledge to call upon than does molecular biology and that fact likewise explains their relative lack of progress.

These two failings – the lack of controlled experiments and of significant relevant outside background knowledge – cause and exacerbate numerous other cognitive obstacles to good social science, many of which we have implicitly identified in earlier chapters. Among those obstacles are:[2]

(1) Failing to investigate *ceteris paribus* clauses in the ways necessary to prove them trustworthy. For example, far too much statistical work in the social sciences rests on *ceteris paribus* assumptions that are often identified only to be then ignored.

(2) The failure to rule out competing hypotheses. Consistency with the data is too often taken to be good proof, when there may nonetheless be several or more competing explanations equally consistent with that data. Statistical practice is again a prime example. As we saw in Chapter 3, a single set of correlations may be compatible with several different causal relations between the variables, and the problem grows exponentially as the number of causal factors increases. Thus finding a model consistent with the data is not enough for good science, though it is often treated as such.

(3) Failing to do fair tests. Recall that a fair test in my terminology is one that provides evidence for a hypothesis without making assumptions that beg the question against the most serious competing hypotheses. Frequently competing social theories are so vaguely formulated that we cannot determine exactly what they do and do not share – thus making fair tests difficult.

(4) Not searching for new sources of data and not acknowledging the weaknesses in existing data. Only so much information can be wrung out of general surveys and the like. Finding tests that will more strongly decide between competing hypotheses often requires finding new data. Statistical practice is again a case in point, for eventually running more and more sophisticated statistical tests on the same data sets reaches a point of diminishing

[2] For a useful discussion along the lines of what follows see Blalock (1989).

returns where convincing evidence will come only from testing competitors against different data. New data can be found – Paige's creative measures of political movements or Hannan and Freeman's measurement of foundings and survivals are obvious examples of what can be done.

(5) Treating correlations as an end in themselves rather than as evidence for causal explanations. Correlations may be useful on their own, but their primary interest lies in inferring causes. Social researchers are too often content simply to report correlations without investigating their causal significance. That practice deprives the research of much of its interest.

(6) The lack of clearly formulated causal claims. Many social science models are formulated in a purely verbal way, and in a manner where the precise causal relations are obscure. That makes fair tests, cross tests, and independent tests difficult. Prime examples in this connection are the grand theorists like Marx, whose theory clearly must have causal implications but who seldom specified detailed predictions about causes.

(7) An overemphasis on grand, highly abstract theory. Parsons comes to mind as a prime example in sociology. Much energy and effort is lost debating a theory that makes few explicit causal claims and has nebulous links to empirical data. The result is wheel spinning rather than cross tests, fair tests, and independent tests leading to well-confirmed causal explanations.

(8) Confusing a simplified, heuristic model with real explanation. This difficulty is closely tied to problems (1) and (2) above. Examples discussed in earlier chapters include Becker's work on the family and Smith's account of Inuit hunting practices. It is all too easy to begin studying a model with no pretensions to explaining actual processes and then subtly shifting to making claims about real causes in the world – without doing the hard work needed to make that move legitimate. This practice is particularly common in economics.[3]

(9) Failing, when offering functional explanations, to consider competing non-functional causes, not showing that persistence is caused by beneficial effects or even correlated with them, not identifying mechanisms in optimality arguments, and other such problems specific to confirming functional explanations.

As I hope is obvious, these failings are interrelated and consist largely of various ways of failing to do fair tests, cross tests, and independent tests

---

[3] See Kincaid (1995) for a number of such moves in economics.

or failing to provide good – that is, causal – explanations; they are also exacerbated by the lack of controlled experiments and relevant background knowledge from other domains. Exactly how important these factors are and to what extent they explain poor social science remain open questions. What we do know – from the work of Paige, Hannan and Freeman, Brown and Harris, and others – is that they are not inevitable and fatal to the prospects for good social science.

We may of course push back the question one step further and ask why these failures are rife in the social sciences. The relative dearth of controlled experiments and background knowledge from other disciplines is part of the explanation. However, we can further explain these failings by looking at the social structure of social science, and it is to hypotheses of this sort that I turn now.

Fundamental, I suspect, to most failures in the social sciences is their very relevance to our lives. Most social research is no more than a few steps removed from moral and political issues of great concern and, correspondingly, direct individual self-interest. For example, we saw in Chapter 5 that holism and individualism have strong moral overtones; yet these are relatively abstract doctrines. The closeness of moral concerns becomes even more apparent when we are discussing the causes of inequality, poverty, educational failure, economic growth, and the like. So powerful forces work against objectivity. Moral concerns can insinuate themselves in many different and often unseen ways into theory assessment, data gathering (or lack thereof), and the other epistemic components of science. As Longino so nicely points out in her work (see Section 2.6), the gap between data and hypothesis leaves room for value assumptions to play a role; the greater that gap – the more we rely on uncertain background assumptions to draw conclusions from the data – the greater the prospects that moral concerns undermine objectivity. Ignoring competing hypotheses, failing to make fair tests, failing to pursue potentially troublesome variables, and the like naturally result when social scientific researchers have a moral ax to grind or depend for their support upon those who do.

Determining exactly how much weak social science comes from normative intrusions would again require a detailed empirical study. But we can note that evidence from the natural sciences supports our hypothesis as well. What areas of natural science research have been the most unsuccessful? Arguably those parts of biology that bear most directly on moral and political interests – sociobiology, evolutionary history of races and gender, and research on sex differences come immediately to mind. These areas have moral and political implications; these areas have to make inferences across a wide gap from limited data to hypotheses. As in the social sciences, these two factors are potent producers of suspect science.

The minuscule resources directed to the social sciences also help explain their slow progress. In 1991, for example, total U.S. government spending for the social sciences was $189 million.[4] Spending for the natural sciences was approximately $12 billion or sixty-three times what was spent on the social sciences. The budget for mathematics, a discipline not in need of data, equaled the combined budgets of anthropology, sociology, economics, and political science. In most years the increase in spending on physical sciences or life sciences is many times greater than the total spent for social science research. Moreover, these are just the disparities in federal resources. Differences in university and industrial spending will exacerbate these differences. At my university, for example, there is no one employed full-time for social science research, but over a thousand are employed full-time for research in the natural sciences.

Numerous other social factors may be partial causes of the poor track record of social research. Making such hypotheses, however, leaves us on thin ice, for we, especially we philosophers, know relatively little about the real social processes of science. Nonetheless, let me suggest that the following are important obstacles in progress:

(1) The reward structures – tenure, promotion, disciplinary stature, and the like – in the social sciences emphasize quantity over quality, thus discouraging the methodological rigor necessary for cross tests, fair tests, and independent tests and their more specific embodiments.

(2) Relatedly, the reward structure does not strongly encourage data gathering, for that is time-consuming work. Individuals who gather data are penalized so long as journals, reviewers, and others let pass work that does a questionable job of measuring the variables at issue, let pass work that does not try to rule out competing explanations, and so on.

(3) There is a lack of cooperative research. This makes rigorous testing of hypotheses about complex phenomena harder – because resources are diminished, because cooperative competition and criticism are promoted by group research, and because data collecting frequently cannot be done by single individuals.

(4) There is a lack of rigorous training in available methods. Natural science students, at both the graduate and undergraduate levels, get far more training in quantitative techniques.

(5) There is a lack of data sharing and other cooperative practices, which makes cross testing – the building on and indirect testing of the results of others – much more difficult.

[4] Natural Science Foundation (1991).

Problems and prospects 265

(6) Students attracted to social science programs tend to dislike quantitative methods.[5]

(7) The book market in the social sciences tends to target a general audience, thus discouraging rigor in research and presentation.

(8) Funds for social science research, compared to natural science research, come more often from entities looking for support for a specific social or political outlook.

(9) Perhaps most obviously, there is a lack of norms promoting the kind of cognitive virtues exemplified by cross tests, fair tests, and independent tests. As we saw in Chapter 5, normative explanations like this one are not deep explanations, but it certainly seems that there are important differences in research traditions between the natural and social sciences that partly explain their differential success. Fleishman and Pons were ridiculed and driven from the ranks of respected scientists when they announced their cold fusion results without doing obvious controls; social scientists do analogous things all the time without raising an eyebrow.

These are not deeply original explanations nor are they obviously right. Yet they do point out the kind of research we need to do to understand why the social sciences do not do better. I should note in this connection that the social constructivist and rhetorical approaches have much to offer here – not for their skeptical conclusions but for what they can tell us about how social science really works.

Implicit in the explanations given above are, of course, recommendations for what the social sciences should be doing. Some of the positive recommendations that flow from the discussion of this chapter and previous ones are: produce clear causal hypotheses and eschew highly abstract theories; produce evidence that implicit *ceteris paribus* assumptions do not undermine the evidence involved; tie causal hypotheses to concrete detail; look for evidence that rules out competing explanations; look for individualist mechanisms while pursuing patterns at the social level; pursue explanations that test and build on the previous work of others; and find novel ways to gather data. These are recommendations that the social sciences can meet. The failings in social research rehearsed in this chapter are contingent facts that can be and sometimes are overcome by the social sciences themselves. So I have argued at length in this book.

---

[5] Here I rely again on Blalock (1989), who reports that Graduate Record Examination scores in the social sciences are lower than natural science averages on both the quantitive *and* verbal components.

REFERENCES

Abel, Theodore. 1948. "The Operation Called *Verstehen.*" *American Journal of Sociology* 54: 211–218.
Achinstein, Peter. 1980. *The Nature of Explanation.* Oxford: Oxford University Press.
——— 1985. "The Method of Hypothesis." In *Observation, Experiment and Hypothesis in Modern Physical Science,* pp. 127–147. See Achinstein and Hanaway (1985).
Achinstein, Peter, and Hanaway, Owen, eds. 1985. *Observation, Experiment and Hypothesis in Modern Physical Science.* Cambridge, Mass.: MIT Press.
Agassi, Joseph. 1973. "Methodological Individualism." In *Modes of Individualism and Collectivism,* pp. 183–214. See O'Neill (1973).
Akerloff, George. 1970. "The Markets for Lemons: Quality, Uncertainty and the Market." *Quarterly Journal of Economics* 84: 488–500.
Alexander, J., ed. 1985. *Neofunctionalism.* Beverly Hills, Calif.: Sage Publications.
Allen, M. R. 1967. *Male Cults and Secret Initiations in Malanesia.* Melbourne, Australia: Melbourne University Press.
Arrow, K., and Debreu, G. 1954. "Existence of Equilibrium for a Competitive Economy." *Econometrica* 20: 265–290.
Arrow, K., and Hahn, F. 1971. *General Competitive Analysis.* San Francisco: Holden Day.
Askari, Hossein, and Cummings, John. 1977. "Estimating Agricultural Supply Response with the Nerlove Model: A Survey." *International Economic Review* 18: 257–293.
Asquith, P., and Hacking, I., eds. 1981. *PSA 1978,* vol. 2. East Lansing, Mich.: Philosophy of Science Association.
Asquith, P., and Nickles, T., eds. 1983. *PSA 1982,* vol. 2. East Lansing, Mich.: Philosophy of Science Association.
Barnes, S. 1982. *T. S. Kuhn and Social Science.* New York: Columbia University Press.
Barrett, R., and Gibson, R., eds. 1990. *Perspectives on Quine.* Cambridge, England: Basil Blackwell.
Barry, Brian. 1978. *Sociologists, Economists, and Democracy.* Chicago: University of Chicago Press.
Barten, A. 1967. "Evidence on the Slutsky Conditions for Demand Equations." *Review of Economics and Statistics* 49: 77–84.
Barten, A., and Geyskens, E. 1975. "The Negativity Condition in Consumer Demand." *European Economic Review* 6: 227–260.

References

Bates, F., and Harvey, C. 1975. *The Structure of Social Systems.* New York: Macmillan.
Becker, Gary. 1976a. *The Economic Approach to Human Behavior.* Chicago: University of Chicago Press.
   1976b. "Irrational Behavior and Economic Theory." In *The Economic Approach to Human Behavior,* pp. 151–169. See Becker (1976a).
   1981. *A Treatise on the Family.* Cambridge, Mass.: Harvard University Press.
Becker, G., Grossman, M., and Murphy, K. 1991. "Rational Addiction and the Effect of Price on Consumption." *American Economic Review* 81: 237–242.
Becker, Morton, 1959. *The Biological Way of Thought.* New York: Columbia University Press.
Begg, David. 1982. *The Rational Expectations Revolution in Macroeconomics.* Oxford: Philip Allan.
Berryman, Alan. 1981. *Population Systems.* New York: Plenum Press.
Bigelow, John, and Pargetter, Robert. 1987. "Functions." *Journal of Philosophy* 84: 181–196.
Blalock, Hubert. 1989. "The Real and Unrealized Contributions of Quantitative Sociology." *American Sociological Review* 54: 447–461.
Blalock, Hubert, ed. 1971. *Causal Models in the Social Sciences.* Chicago: Atherton.
Blaug, Mark. 1976. "The Empirical Status of Human Capital Theory: A Slightly Jaundiced Survey." *Journal of Economic Literature* 14: 827–855.
Bloor, David. 1976. *Knowledge and Social Imagery.* London: Routledge and Kegan Paul.
Blundell, Richard, Pashardes, Panos, and Weber, G. 1993. "What Do We Learn About Consumer Demand Patterns From Micro Data?" *American Economic Review* 83: 570–598.
Boland, Lawrence. 1982. *The Foundations of Economic Method.* Boston: Allen and Unwin.
Borse, Christopher. 1984. "Wright on Functions." In *Conceptual Issues in Evolutionary Biology,* pp. 369–386. See Sober, ed. (1984).
Bowles, S., and Gintis, H. 1971. *Schooling in Capitalist America.* New York: Basic Books.
Boyd, R., and Richerson, P. 1985. *Culture and the Evolutionary Process.* Chicago: University of Chicago Press.
Bronsard, Camille, and Salvas-Bronsard, Lise. 1984. "On Price Exogeneity in Complete Demand Systems." *Journal of Econometrics* 24: 235–247.
Brown, Brendan. 1989. *The Economics of Swap Markets.* London: Routledge.
Brown, George W., and Harris, Tirril. 1978. *Social Origins of Depression.* New York: Free Press.
Brown, James. 1989. *The Rational and the Social.* London: Routledge.
Campbell, Donald T. 1975. "Degress of Freedom and the Case Study." *Comparative Political Studies* 8: 178–193.
Carnap, Rudolf. 1934. *The Unity of Science,* trans. M. Black. London: K. Paul, Trench, and Trubner.
   1950. *Logical Foundations of Probability.* London: Routledge and Kegan Paul.
Carroll, Glenn. 1987. *Publish and Perish: The Organizational Ecology of Newspaper Industries.* Greenwich, Conn.: JAI Press.
Cartwright, Nancy. 1984. *How the Laws of Physics Lie.* New York: Oxford University Press.
   1989. *Nature's Capacities and Their Measurement.* Oxford: Clarendon Press.
   1993 "How We Relate Theory to Observation." In *World Changes,* pp. 259–275. See Horwich (1993).
Cassel, G. 1932. *The Theory of Social Economy,* trans. S. Barron. New York: Harcourt Brace.
Causey, Robert. 1977. *Unity of Science.* Boston: D. Reidel.
Cheney, Dorothy, and Seyfarth, Robert. 1990. *How Monkeys See the World: Inside the Mind of Another Species.* Chicago: University of Chicago Press.

Churchland, Paul. 1979. *Scientific Realism and the Plasticity of Mind.* Cambridge University Press.
Cohen, G. A. 1978. *Karl Marx's Theory of History: A Defence.* Princeton: Princeton University Press.
Cole, Stephen. 1992. *Making Science: Between Nature and Society.* Cambridge, Mass.: Harvard University Press.
Coleman, James. 1990. *Foundations of Social Theory.* Cambridge, Mass.: Harvard University Press.
Collins, Randall. 1981. "On the Microfoundations of Macrosociology." *American Journal of Sociology* 86: 984–1014.
  1988. *Theoretical Sociology.* Fort Worth, Tex.: Harcourt and Brace.
Cummins, Robert. 1975. "Functional Analysis." *Journal of Philosophy* 72: 741–764.
Damuth, John, and Heiser, Loraine. 1988. "Alternative Formulations of Multilevel Selections." *Biology and Philosophy* 3: 407–431.
Davidson, Donald. 1970. "Mental Events." In *Experience and Theory,* pp. 79–101. See Foster and Swanson (1970).
  1980. *Essays on Actions and Events.* Oxford: Oxford University Press.
Davis, Kingsley, and Moore, Stanley. 1945. "Some Principles of Stratification." *American Sociological Review* 10: 242–247.
Day, Timothy, and Kincaid, Harold. 1994. "Putting Inference to the Best Explanation in Its Place." *Synthese* 98: 271–295.
Deaton, Angus. 1972. "Surveys in Applied Economics: Models of Consumer Behavior." *Economic Journal* 328: 1145–1236.
  1974. "The Analysis of Consumer Demand in the United Kingdom, 1900–1970." *Econometrica* 42: 341–367.
Deaton, Angus, and Muellbauer, John. 1980. *Economics and Consumer Behavior.* Cambridge University Press.
de Jasay, Anthony. 1985. *The State.* Oxford: Basil Blackwell.
de Marchi, Neil, ed. 1988. *The Popperian Legacy in Economics.* Cambridge University Press.
de Marchi, Neil, and Blaug, Mark, eds. 1991. *Appraising Economic Theories.* Worcester, England: Edward Elger.
Dickman, Joel. 1990. "Two Qualms About Functionalist Marxism." *Philosophy of Science* 57: 631–643.
Dilthey, Wilhelm. 1989. *Introduction to the Human Sciences.* Princeton: Princeton University Press.
Dumhoff, 1967. *Who Rules America?* Englewood Cliffs, N.J.: Prentice Hall.
Doppelt, Gerald. 1988. "The Philosophic Requirements for an Adequate Conception of Scientific Rationality." *Philosophy of Science* 55: 104–34.
  1990. "The Naturalist Conception of Methodological Standards in Science." *Philosophy of Science* 57: 1–20.
Downs, Anthony. 1957. *An Economic Theory of Democracy.* New York: Harper.
Duhem, Pierre. 1954. *The Aim and Structure of Physical Theory.* Princeton: Princeton University Press.
Dupre, John. 1987. *The Latest on the Best: Essays on Evolution and Optimality.* Cambridge, Mass.: MIT Press.
Durkheim, Emile. 1933. *The Division of Labor in Society.* New York: Macmillan.
  1965. *The Rules of the Sociological Method.* New York: Free Press.
Earman, John. 1978. "The Universality of Laws." *Philosophy of Science* 45: 173–181.
Earman, John, ed. 1992. *Inference, Explanation, and Other Frustrations: Essays in the Philosophy of Science.* Berkeley: University of California Press.

Ellen, R. 1982. *Environment, Subsistence and System: The Ecology of Small Scale Social Formations.* Cambridge University Press.
Elster, Jon. 1983. *Explaining Technical Change.* Cambridge University Press.
— 1985. *Making Sense of Marx.* Cambridge University Press.
— 1989. *Nuts and Bolts for the Social Sciences.* Cambridge University Press.
Endler, John. 1986. *Natural Selection in the Wild.* Princeton: Princeton University Press.
Faia, M. 1986. *Dynamic Functionalism: Strategy and Tactics.* Cambridge University Press.
Fay, Brian. 1984. "Naturalism as a Philosophy of Social Science." *Philosophy of the Social Sciences* 14: 529–542.
Feleppa, Robert. 1988. *Convention, Translation and Understanding: Philosophical Problems in the Comparative Study of Culture.* Albany, N.Y.: SUNY Press.
Feynman, Richard. 1965. *The Character of Physical Law.* Cambridge, Mass.: MIT Press.
Fine, Arthur, Forbes, M., and Wessels, L., eds. 1990. *PSA 1990,* vol. 1. East Lansing, Mich.: Philosophy of Science Association.
Fine, Arthur, and Leplin, Jarrett, eds. 1988. *PSA 1988,* vol. 1. East Lansing, Mich.: Philosophy of Science Association.
Fisk, Milton. 1989. *The State and Justice: An Essay in Political Theory.* Cambridge University Press.
Fodor, J. A. 1974. "Special Sciences (Or: The Disunity of Science as a Working Hypothesis)." *Synthese* 28: 97–115.
Fodor, Jerry, and LePore, Ernest. 1992. *Holism: A Shopper's Guide.* Oxford: Basil Blackwell.
Foster, L., and Swanson, J., eds. 1970. *Experience and Theory.* Amherst: University of Massachusetts Press.
Franklin, Allan. 1984. "Are Paradigms Incommensurable?" *British Journal for Philosophy of Science* 35: 57–60.
— 1986. *The Neglect of Experiment.* Cambridge University Press.
— 1990. *Experiment Right or Wrong.* Cambridge University Press.
Freeman, Derek. 1983. *Margaret Mead and Samoa.* New York: Penguin.
Friedman, Michael. 1983. *Foundations of Space-Time Theories: Relativistic Physics and Philosophy of Science.* Princeton: Princeton University Press.
— 1992. "Philosophy and the Exact Sciences: Logical Positivism as a Case Study." In *Inference, Explanation, and Other Frustrations: Essays in the Philosophy of Science,* pp. 84–99. See Earman (1992).
Friedman, Milton. 1953a. *Essays in Positive Economics.* Chicago: University of Chicago Press.
— 1953b. "The Methodology of Positive Economics." In *Essays in Positive Economics,* pp. 3–43. See Milton Friedman (1953a).
Gadamer, Hans-Georg. 1977. *Philosophical Hermeneutics,* trans. D. Linge. Berkeley: University of California Press.
Galison, Peter. 1987. *How Experiments End.* Chicago: University of Chicago Press.
Garber, Daniel. 1986. "Learning from the Past." *Synthese* 67: 91–114.
Garfinkel, Alan. 1981. *Forms of Explanation.* New Haven, Conn.: Yale University Press.
Geertz, Clifford. 1973a. "Deep Play: Notes on the Balinese Cockfight." In *The Interpretation of Cultures,* pp. 412–455. See Geertz (1973b).
— 1973b. *The Interpretation of Cultures.* New York: Basic Books.
Geertz, Hildred, and Geertz, Clifford. 1973. *Kinship in Bali.* Chicago: University of Chicago Press.
Gibson, Quinton. 1983. "Tendencies." *Philosophy of Science* 50: 296–309.
Giddens, Anthony. 1976. *New Rules of the Sociological Method.* New York: Basic Books.
— 1979. *Central Problems in Social Theory.* Berkeley: University of California Press.

Giere, Ron. 1988. *Explaining Science: A Cognitive Approach.* Chicago: University of Chicago Press.
Gilbert, Christopher. 1991. "Do Economists Test Theories? – Demand Analysis and Consumption Analysis as Tests of Theories of Economic Methodology." In *Appraising Economic Theories,* pp. 137–168. See de Marchi and Blaug (1991).
Glymour, Clark. 1980. *Theory and Evidence.* Princeton: Princeton University Press.
Glymour, Clark, Scheines, Richard, Spirtes, Peter, and Kelly, Kevin. 1987. *Discovering Causal Structure.* Orlando, Fla.: Academic Press.
Goldfield, S., and Quandt, R. 1981. "Single Market Disequilibrium Models: Estimation and Testing." *Journal of Econometrics* 13: 157.
Goldman, Alvin. 1986. *Epistemology and Cognition.* Cambridge, Mass.: Harvard University Press.
   1987. "Foundations of Social Epistemics." *Synthese* 73: 109–144.
Goldman, Alvin, and Shaked, Moshe. 1991. "An Economic Model of Scientific Activity and Truth Acquisition." *Philosophical Studies* 63: 31–55.
Goodman, Nelson. 1965. *Fact, Fiction and Forecast.* Indianapolis: Bobbs Merrill.
Grant, Peter. 1986. *Ecology and Evolution of Darwin's Finches.* Princeton: Princeton University Press.
Greenwood, John. 1991. *Relations and Representation: An Introduction to the Philosophy of Psychological Science.* London: Routledge.
Guttenplan, S., ed. 1975. *Mind and Language.* Oxford: Clarendon Press.
Habermas, Jurgen. 1985. *Theory of Communicative Action,* trans. T. McCarthy. Boston: Beacon Press.
Hacking, Ian. 1993 "Working in a New World: The Taxonomic Solution." In *World Changes,* pp. 275–311. See Horwich (1993).
Hahn, L., and Schlipp, P., eds. 1986. *The Philosophy of W. V. Quine.* LaSalle, Ill.: Open Court.
Hallpike, C. 1986. *Principles of Social Evolution.* New York: Oxford University Press.
Hands, Wade. 1993. *Testing, Rationality and Progress.* Lanham, Md.: Rowman and Littlefield.
Hannan, M., and Freeman, J. 1989. *Organizational Ecology.* Cambridge, Mass.: Harvard University Press.
Harman, Gilbert. 1965. "Inference to the Best Explanation." *Philosophical Review* 74: 88–95.
   1986. *Change in View.* Cambridge University Press.
Harris, Marvin. 1979. *Cannibals and Kings.* New York: Random House.
   1985. *Good to Eat.* New York: Simon and Schuster.
Haugeland, J., ed. 1981. *Mind Design.* Cambridge, Mass.: MIT Press.
Hausman, Daniel. 1981a. "Are General Equilibrium Theories Explanatory?" In *Philosophy in Economics,* pp. 17–32. See Pitt (1981).
   1981b. *Capital, Profits and Prices.* New York: Columbia University Press.
   1984. "Defending Microeconomic Theory." *Philosophical Forum* 15: 392–404.
   1992. *The Inexact and Separate Science of Economics.* Cambridge University Press.
Hechter, M. 1983. *The Microfoundations of Macrosociology.* Philadelphia: Temple University Press.
Hellman, Geoffrey, 1979. "Historical Materialism." In *Issues in Marxist Philosophy,* pp. 143–170. See Mepham and Ruben (1979).
   "Realist Principles." *Philosophy of Science* 50: 227–249.
Hellman, Geoffrey, and Thompson, F. W. 1975. "Physicalism: Ontology, Determination and Reduction." *Journal of Philosophy* 72: 551–564.
Hempel, Carl. 1965. *Aspects of Scientific Explanation and Other Essays in the Philosophy of Science.* New York: Free Press.

1988. "Provisoes: A Problem Concerning the Inferential Function of Scientific Theories." *Erkenntnis* 28: 147–164.
Henderson, David. 1991. "On the Testability of Psychological Generalizations." *Philosophy of Science* 58: 586–607.
1993. *Interpretation and Explanation in the Human Sciences.* Albany, N.Y.: SUNY Press.
Hey, John D., ed. 1989. *Current Issues in Microeconomics.* New York: St. Martin's Press.
Hindness, Barry. 1988. *Choice, Rationality, and Social Theory.* London: Unwin Hyman.
Hirsch, Abraham, and de Marchi, Neil. 1990. *Milton Friedman: Economics in Theory and Practice.* Ann Arbor: University of Michigan Press.
Homans, George. 1974. *Social Behavior: Its Elementary Forms.* New York: Harcourt Brace Jovanovich.
Homans, George, and Schneider, David. 1955. *Marriage, Authority and Final Causes.* New York: Free Press.
Hookaway, Christopher. 1988. *Quine.* Stanford, Calif.: Stanford University Press.
Horan, Barbara. 1990. "What Price Optimality?" *Biology and Philosophy* 5: 105–124.
Horgan, Terence, and Graham, George. 1991. "In Defense of Southern Fundamentalism." *Philosophical Studies* 62: 107–134.
Horwich, Paul, ed. 1993. *World Changes.* Cambridge, Mass.: MIT Press.
Howson, Colin, and Urbach, Peter. 1989. *Scientific Reasoning: The Bayesian Approach.* La-Salle, Ill.: Open Court.
James, Susan. 1984. *The Content of Social Explanation.* Cambridge University Press.
Janssen, Maarten. 1993. *Microfoundations: A Critical Inquiry.* London: Routledge.
Johnson, Chalmers. 1966. *Revolutionary Change.* Boston: Little, Brown.
Keesing, Roger. 1987. "Anthropology as an Interpretive Quest." *Current Anthropology* 28: 162–176.
Kim, Jaegwon. 1989. "The Myth of Non-reductive Materialism." *Proceedings and Addresses of the American Philosophical Association* 63: 31–47.
Kincaid, Harold. 1986. "Reduction, Explanation, and Individualism." *Philosophy of Science* 53: 492–513.
1987. "Supervenience Doesn't Entail Reducibility." *Southern Journal of Philosophy* 25: 342–356.
1988. "Supervenience and Explanation." *Synthese* 77: 251–281.
1990a. "Assessing Functional Explanation in the Social Sciences." In *PSA 1990*, vol. 1, p. 341–354. See Fine et al. (1990).
1990b. "Defending Laws in the Social Sciences." *Philosophy of the Social Sciences* 20: 56–83.
1990c. "Discussion: Eliminativism and Methodological Individualism." *Philosophy of Science* 57: 141–148.
1990d. "Molecular Biology and the Unity of Science." *Philosophy of Science* 57: 575–593.
1995. "Optimality Arguments and the Theory of the Firm." In *On the Reliability of Economic Models.* See Little (1995).
in press a. *Individualism and the Unity of Science: Essays on Reduction, Explanation, and the Special Sciences.* Lanham, Md.: Rowman and Littlefield.
in press b. "The Empirical Presuppositions of Metaphysical Explanations in Economics." *The Monist.*
Kirk, Robert. 1986. *Translation Determined.* Oxford: Clarendon Press.
Kitcher, Philip. 1978. "Theories, Theorists and Theoretical Change." *Philosophical Review* 87: 519–547.

1984. "Species." *Philosophy of Science* 51: 308–333.
1985. *Vaulting Ambition.* Cambridge, Mass.: MIT Press.
1987. "Why Not the Best." In *The Latest on the Best: Essays on Evolution and Optimality*, pp. 77–102. See Dupre (1987).
1989. "Explanatory Unification and the Causal Structure of the World." In *Scientific Explanation*, pp. 410–507. See Kitcher and Salmon (1989).
1993. *The Advancement of Science: Science Without Legend, Objectivity Without Illusions.* New York: Oxford University Press.
Kitcher, Philip, and Salmon, Wesley, eds. 1989. *Scientific Explanation.* Minneapolis: University of Minnesota Press.
Kornblith, Hilary, ed. 1985. *Naturalizing Epistemology.* Cambridge, Mass.: MIT Press.
Korsch, Karl. 1938. *Karl Marx.* London: Russell and Russell.
Kosso, Peter. 1989. *Observability and Observation in Physical Science.* Dordrecht, Netherlands: Kluwer.
Kreps, David. 1990. *Game Theory and Economic Modelling.* Oxford: Oxford University Press.
Kuhn, Thomas. 1962. *The Structure of Scientific Revolutions* (first edition). Chicago: University of Chicago Press.
1970. *The Structure of Scientific Revolutions* (second edition). Chicago: University of Chicago Press.
1974a. *The Essential Tension.* Chicago: University of Chicago Press.
1974b. "Objectivity, Value Judgment, and Theory Choice." In *The Essential Tension*, pp. 320–339. See Kuhn (1974a).
Kuper, Adam. 1977. *The Social Anthropology of Radcliffe-Brown.* London: Routledge and Kegan Paul.
Lakatos, Imre. 1970. "Falsification and the Methodology of Scientific Research Programmes." In *Criticism and the Growth of Knowledge*, pp. 91–196. See Lakatos and Musgrave (1970).
Lakatos, Imre, and Musgrave, Alan., eds. 1970. *Criticism and the Growth of Knowledge.* Cambridge University Press.
Langbein, Laura, and Lichtman, Allan. 1978. *Ecological Inference.* Beverly Hills, Calif.: Sage Publications.
Latour, Bruno. 1987. *Science in Action: How to Follow Scientists and Engineers Through Society.* Cambridge, Mass.: Harvard University Press.
Latour, Bruno, and Woolgar, Steve. 1979. *Laboratory Life: The Construction of Scientific Facts.* Princeton: Princeton University Press.
Laudan, Larry. 1984. *Science and Values: The Aims of Science and Their Role in Scientific Debate.* Berkeley: University of California Press.
1987. "Progress or Rationality? The Prospects for Normative Naturalism." *American Philosophical Quarterly* 24: 19–31.
1990. "Normative Naturalism." *Philosophy of Science* 57: 60–78.
Laudan, Larry, and Leplin, Jarret. 1991. "Empirical Equivalence and Underdetermination." *Journal of Philosophy* 88: 449–473.
Laudan, Rachel. 1981. "The Recent Revolution in Geology and Kuhn's Theory of Scientific Change." In *PSA 1978*, vol. 2, pp. 227–239. See Asquith and Hacking (1981).
Laudan, Rachael, and Laudan, Larry. 1989. "Dominance and the Disunity of Method: Solving the Problems of Innovation and Consensus." *Philosophy of Science* 56: 221–237.
Leach, E., ed. 1967. *The Structural Study of Myth and Totemism.* London: Tavistock.
Leamer, Edward. 1983. "Let's Take the Con Out of Econometrics." *American Economic Review* 73: 31–43.

References

Lenoir, Timothy. 1982. *Strategies of Life*. Chicago: University of Chicago Press.
Leplin, Jarrett. 1990. "Renormalizing Naturalism." *Philosophy of Science* 57: 20–34.
Lessnoff, Michael. 1974. *The Structure of Social Science*. London: Allen and Unwin.
Levins, Richard. 1966. "The Strategy of Model Building in Population Biology." *American Scientist* 54: 421–431.
— 1968. *Evolution in Changing Environments*. Princeton: Princeton University Press.
Levinson, David, and Malone, Martin. 1984. *Toward Explaining Human Culture*. New Haven, Conn.: Yale University Press.
Lévi-Strauss, Claude. 1967. "The Story of Asdiwal." In *The Structural Study of Myth and Totemism*, pp. 1–47. See Leach (1967).
Lewis, David. 1969. *Convention*. Cambridge, Mass.: Harvard University Press.
Lewontin, R. C. 1963. "Models, Mathematics, and Metaphors." *Synthese* 15: 222–244.
Lieberson, Stanley. 1992. "Some Thoughts About Evidence in Sociology." *American Sociological Review* 57: 1–15.
Liska, Allen E. 1975. *The Consistency Controversy: Readings on the Impact of Attitude on Behavior*. New York: Wiley and Sons.
Little, Daniel. 1989. *Understanding Peasant China*. New Haven, Conn.: Yale University Press.
— 1991. *Varieties of Social Explanation: An Introduction to the Philosophy of Social Science*. Boulder, Colo.: Westview Press.
Little, Daniel, ed. 1995. *On the Reliability of Economic Models*. Dordrecht, Netherlands: Kluwer.
Longino, Helen. 1990. *Science as Social Knowledge*. Princeton: Princeton University Press.
Loomes, Graham. 1989. "Experimental Economics." In *Current Issues in Microeconomics*, pp. 152–179. See Hey (1989).
Maddox, John. 1992. "Is Molecular Biology Yet a Science?" *Nature* 355: 201.
Maher, Patrick. 1988. "Prediction, Accommodation, and the Logic of Discovery." In *PSA 1988*, vol. 1, pp. 273–286. See Fine and Leplin (1988).
Marsh, Catherine. 1982. *The Survey Method: The Contribution of Surveys to Sociological Explanation*. London: Allen and Unwin.
Martin, Raymond. 1989. *The Past Within Us*. Princeton: Princeton University Press.
Marwell, G., and Oliver, P. 1993. *The Critical Mass in Collective Action: A Micro-Social Theory*. Cambridge University Press.
Mathien, Thomas. 1988. "Network Analysis and Methodological Individualism." *Philosophy of Social Science* 18: 1–20.
Mayhew, Bruce. 1980. "Structuralism Versus Individualism: Part 1, Shadowboxing in the Dark." *Social Forces* 59: 335–375.
Maynard Smith, J., Burian, R., Kaufman, S., Alberch, P., Campbell, J., Goodwin, B., Lande, R., Raup, D., and Wolpert, L.. 1985. "Developmental Constraints and Evolution." *Quarterly Review of Biology* 60: 265–286.
McCarthy, Mark. 1975. "On Methodological Individualism." Diss., Indiana University.
McCloskey, Donald. 1985. *The Rhetoric of Economics*. Madison: University of Wisconsin Press.
McDonald, Graham, and Petit, Philip. 1981. *Semantics and Social Science*. London: Routledge and Kegan Paul.
McMullin, Ernan. 1982. "Values in Science." In *PSA 1982*, vol. 2, pp. 3–28. See Asquith and Nickles (1983).
Mepham, D., and Ruben, D.-H., eds. 1979. *Issues in Marxist Philosophy*. Atlantic Highlands, N.J.: Humanities Press.
Miliband, Ralph. 1969. *The State in Capitalist Society*. New York: Basic Books.

Miller, Richard. 1978. "Methodological Individualism and Social Explanation." *Philosophy of Science* 45: 387–414.
    1987. *Fact and Method.* Princeton: Princeton University Press.
Mirowski, Philip. 1989. *More Heat Than Light: Economics as Social Physics, Physics as Nature's Economics.* Cambridge University Press.
    1994. "Scissors, Lies and Scotchtape: Lectures on the History of the 'Laws' of Demand and Supply." Typescript shown to author.
Moore, H. L. 1914. *Economic Cycles: Their Law and Cause.* New York: Macmillan.
Myrdal, Gunnar. 1970. *Objectivity in Social Science.* London: Duckworth.
Nagel, Ernest. 1979. *Teleology Revisited.* New York: Columbia University Press.
Natural Science Foundation. 1991. *Federal Funds for Research and Development: Fiscal Years 1989, 1990, 1991,* vol. 34.
Nelson, Alan. 1984. "Some Issues Surrounding the Reduction of Macroeconomics to Microeconomics." *Philosophy of Science* 51: 573–594.
    1989. "Average Explanation." *Erkenntnis* 30: 23–42.
    1990. "Are Economic Kinds Natural Kinds?" In *Scientific Theories,* pp. 102–136. See Savage (1990).
Nelson, John, ed. 1987. *The Rhetoric of the Human Sciences.* Madison: University of Wisconsin Press.
Nelson, Richard, and Winter, Sidney. 1982. *An Evolutionary Theory of Economic Change.* Cambridge, Mass.: Harvard University Press.
Nerlove, Marc. 1979. "The Dynamics of Supply: Retrospect and Prospect." *American Journal of Agricultural Economics* 61: 874–888.
Nickels, T., ed. 1980. *Scientific Discovery.* Boston: D. Reidel.
North, Douglas. 1981. *Structure and Change in Economic History.* New York: Norton.
O'Neill, J., ed. 1973. *Modes of Individualism and Collectivism.* London: Heinemann.
Paige, Jeffrey. 1975. *Agrarian Revolutions.* New York: Free Press.
Papineau, David. 1978. *For Science in the Social Sciences.* New York: St. Martin's Press.
    1991. "Correlations and Causes." *British Journal for the Philosophy of Science* 42: 397–413.
Pavel, C., and Binkey, P. 1987. "Cost and Competition in Bank Credit Cards." *Economic Perspectives* 11: 3–13.
Pearce, David. 1987. *Roads to Commensurability.* Boston: D. Reidel.
Pickering, Andrew. 1990. "Knowledge, Practice, and Mere Construction." *Social Studies of Science* 20: 682–729.
Pitt, J., ed. 1981. *Philosophy in Economics.* Dordrecht, Netherlands: D. Reidel.
Piven, F., and Cloward, R. 1971. *Regulating the Poor: The Functions of Public Welfare.* New York: Pantheon Books.
Pollak, Robert, and Wales, Terence. 1992. *Demand System Specification and Estimation.* New York: Oxford University Press.
Popper, Karl. 1962. *Conjecture and Refutations.* New York: Basic Books.
    1982. *The Open Society and Its Enemies.* London: Routledge and Kegan Paul.
Putnam, Hilary. 1981. "Reductionism and the Nature of Psychology." In *Mind Design,* pp. 205–219. See Haugeland (1981).
Quandt, Richard. 1988. *The Econometrics of Disequilibrium.* London: Basil Blackwell.
Quine, W. V. O. 1953. "Two Dogmas of Empiricism." In *From a Logical Point of View,* pp. 20–47. See Quine (1961).
    1960. *Word and Object.* Cambridge, Mass.: MIT Press.
    1961. *From a Logical Point of View.* New York: Harper and Row.
    1975. "Mind and Verbal Disposition." In *Mind and Language,* pp. 83–95. See Guttenplan (1975).

References

1990. "Comment on Katz." In *Perspectives on Quine*, pp. 198–200. See Barrett and Gibson (1990).
Quine, W. V. O., and Ullian, J. S. 1970. *The Web of Belief.* New York: Random House.
Radcliffe-Brown, A. R. 1977. "The Interpretation of Andaman Island Ceremonies." In *The Social Anthropology of Radcliffe-Brown.* See Kuper (1977).
Rappaport, R. 1984. *Pigs for the Ancestors: Ritual in the Ecology of a New Guinea People.* New Haven, Conn.: Yale University Press.
Redhead, Michael. 1987. *Incompleteness, Nonlocality and Realism.* Oxford: Oxford University Press.
Richardson, Robert. 1979. "Functionalism and Reduction." *Philosophy of Science* 46: 533–558.
Rorty, Richard. 1979. *Philosophy and the Mirror of Nature.* Princeton: Princeton University Press.
1982a. *The Consequences of Pragmatism.* Minneapolis: University of Minnesota Press.
1982b. "Method, Social Science and Social Hope." In *The Consequences of Pragmatism,* pp. 191–211. See Rorty (1982a).
1987. "Science as Solidarity." In *The Rhetoric of the Human Sciences.* See Nelson (1987).
Rosenberg, Alexander, 1976. *Microeconomic Laws.* London: University of Pittsburgh Press.
1978. "The Supervenience of Biological Concepts." *Philosophy of Science* 45: 368–386.
1980. *Sociobiology and the Preemption of Social Science.* Baltimore: Johns Hopkins University Press.
1981. "A Skeptical History of Microeconomic Theory." In *Philosophy in Economics,* pp. 47–63. See Pitt (1981).
1983. "If Economics Isn't a Science, What Is It?" *Philosophical Forum* 14: 296–314.
1985. *The Structure of Biological Science.* Cambridge University Press.
1988. *Philosophy of Social Science.* Boulder, Colo.: Westview Press.
1989. "Are Generic Predictions Enough?" *Erkenntnis* 30: 43–68.
1990. "Normative Naturalism and the Role of Philosophy." *Philosophy of Science* 57: 34–44.
1992. *Economics - Mathematical Politics or Science of Diminishing Returns.* Chicago: University of Chicago Press.
Roth, Alvin. 1991. "A Natural Experiment in the Organization of Entry Level Labor Markets." *American Economic Review* 81: 415–441.
Roth, Paul. 1986a. "Comment on Feleppa." *Current Anthropology* 27: 251–252.
1986b. "Semantics Without Foundations." In *The Philosophy of W. V. Quine,* pp. 433–459. See Hahn and Schlipp (1986).
1987. *Meaning and Method in the Social Sciences.* Ithaca, N.Y.: Cornell University Press.
Rouse, Joseph. 1987. *Knowledge and Power: Toward a Political Philosophy of Science.* Ithaca, N.Y.: Cornell University Press.
Ruben, David. 1985. *The Metaphysics of the Social World.* London: Routledge and Kegan Paul.
Rudner, Richard. 1953. "The Scientist Qua Scientist Makes Value Judgments." *Philosophy of Science* 20: 1–6.
Ryan, Richard. 1991. "Fact, Phenomena, and Fantasies." *National Review.* March 18: 56–58.
Salmon, Merilee. 1990. "On the Possibility of Lawful Explanation in Archaeology." *Critica* 66: 87–114.
1992. "Philosophy of the Social Sciences." In *Introduction to the Philosophy of the Social Sciences.* See Salmon et al. (1992).
Salmon, M., Earman, J., Glymour, C., Lennox, J., Machamer, P., McGuire, J., Norton, J.,

Salmon, W., and Schaffner, K., eds. 1992. *Introduction to the Philosophy of Science.* Englewood Cliffs, N.J.: Prentice Hall.

Salmon, Wesley. 1984. *Scientific Explanation and the Causal Structure of the World.* Princeton: Princeton University Press.

——— 1989. "Four Decades of Scientific Explanation." In *Scientific Explanation,* pp. 3–220. See Kitcher and Salmon (1989).

Savage, Wade, ed. 1990. *Scientific Theories.* Minneapolis: University of Minnesota Press.

Schiller, Bradley. 1984. *The Economics of Poverty and Discrimination.* Englewood Cliffs, N.J.: Prentice Hall.

Schmitt, Frederick, ed. 1994. *Socializing Epistemology: The Social Dimensions of Knowledge.* Lanham, Md.: Rowman and Littlefield.

Schrader, David. 1993. *The Corporation As Anomaly.* Cambridge University Press.

Schultz, H. 1938. *The Theory and Measurement of Demand.* Chicago: University of Chicago Press.

Searle, John. 1984. *Minds, Brains and Behavior.* Cambridge, Mass.: Harvard University Press.

Sensat, Julius. 1988. "Methodological Individualism and Marxism." *Economics and Philosophy* 4: 189–221.

Shapere, Dudley. 1964. "The Structure of Scientific Revolutions." *Philosophical Review* 73: 383–394.

Sheffrin, Steven. 1983. *Rational Expectations.* Cambridge University Press.

Siegel, Harvey. 1985. "What is the Question Concerning the Rationality of Science?" *Philosophy of Science* 52: 517–537.

Simon, H. 1971. "Spurious Correlation: A Causal Interpretation." In *Causal Models in the Social Sciences,* pp. 5–18. See Blalock (1971).

Simoons, F. J. 1979. "Questions in the Sacred-Cow Controversy." *Current Anthropology* 20: 467–493.

Skocpol, Theda. 1979. *States and Social Revolutions.* Cambridge University Press.

Skorupski, John. 1976. *Symbol and Theory.* Cambridge University Press.

Skyrms, Brian. 1980. *Causal Necessity.* New Haven, Conn.: Yale University Press.

Smith, Eric. 1981. "The Application of Optimal Foraging Theory to the Analysis of Hunter-Gatherer Group Size." In *Hunter-Gatherer Foraging Strategies.* See Winterhalder and Smith (1981).

——— 1985. "Inuit Foraging Groups: Some Simple Models Incorporating Conflicts of Interest, Relatedness, and Central-Place Sharing." *Ethology and Sociobiology* 6: 27–47.

——— 1987. "Optimization Theory in Anthropology: Applications and Critiques." In *The Latest on the Best: Essays on Evolution and Optimality,* pp. 201–249. See Dupre (1987).

Sober, Elliott. 1984. *The Nature of Selection.* Cambridge, Mass.: MIT Press.

——— 1988. *Reconstructing the Past.* Cambridge, Mass.: MIT Press.

Sober, Elliott, ed. 1984. *Conceptual Issues in Evolutionary Biology.* Cambridge, Mass.: MIT Press.

Soles, Deborah. 1984. "On the Indeterminacy of Action." *Philosophy of the Social Sciences* 14: 475–488.

Solomon, Miriam. 1994. "A More Social Epistemology." In *Socializing Epistemology,* pp. 217–235. See Schmitt (1994).

Stinchcombe, Arthur. 1968. *Constructing Social Theories.* New York: Harcourt Brace.

Stone, J. R. N. 1954. *The Measurement of Consumer Expenditure and Behavior in the U.K. 1920–38.* Cambridge University Press.

Stroud, Barry. 1985. "The Significance of Naturalized Epistemology." In *Naturalizing Epistemology,* pp. 71–89. See Kornblith (1985).

References

Taylor, Charles. 1971. "Interpretation and the Sciences of Man." *Review of Metaphysics* 25: 3–51.
   1980. "Understanding in Human Science." *Review of Metaphysics* 34: 3–23.
Thagard, Paul. 1988. *Computational Philosophy of Science.* Cambridge, Mass.: MIT Press.
Theil, H. 1975. *Theory and Measurement of Consumer Demand,* vol. 1. Amsterdam: North Holland.
   1976. *Theory and Measurement of Consumer Demand,* vol. 2. Amsterdam: North Holland.
Thomas, L., Kronenfeld, J., and Kronenfeld, D. 1976. "Asdiwal Crumbles: A Critique of Lévi-Straussian Myth Analysis." *American Ethnologist* 3: 147–173.
Toumela, Raimo. 1990. "Discussion: Methodological Individualism and Explanation." *Philosophy of Science* 57: 96–103.
Turner, J., and Maryanski, A. 1979. *Functionalism.* Menlo Park, Calif.: Benjamin-Cummings.
Turner, Victor. 1967. *The Forest of Symbols.* Ithaca, N.Y.: Cornell University Press.
Urbach, Peter. 1985. "Randomization and the Design of Experiments." *Philosophy of Science* 52: 256–273.
van Fraassen, Bas. 1981. *The Scientific Image.* Oxford: Oxford University Press.
van Parijs, P. 1981. *Evolutionary Explanation in the Social Sciences.* Totowa, N.J.: Rowman and Littlefield.
Vayda, Andrew. 1974. "Warfare in Ecological Perspective." *Annual Review of Ecology and Systematics* 5: 183–93.
   1987. "Explaining What People Eat: A Review Article." *Human Ecology* 15: 493–510.
   1989. "Explaining Why Marings Fought." *Journal of Anthropological Research* 45: 159–177.
von Neumann, J. 1945. "A Model of General Economic Equilibrium." *Review of Economic Studies* 13: 1–9.
Watkins, John. 1973. "Methodological Individualism: A Reply." In *Modes of Individualism and Collectivism,* pp. 179–185. See O'Neill (1973).
Weintraub, Roy. 1979. *Microfoundations: The Compatibility of Microeconomics and Macroeconomics.* Cambridge University Press.
   1985. *General Equilibrium Analysis.* Cambridge University Press.
   1988. "The Neo-Walrasian Program Is Empirically Progressive." In *The Popperian Legacy,* pp. 213–230. See de Marchi (1988).
   1991. *Stabilizing Dynamics.* Cambridge University Press.
Will, Clifford. 1986. *Was Einstein Right?* New York: Basic Books.
Williamson, Oliver. 1975. *Markets and Hierarchies: Analysis and Antitrust Implications.* New York: Free Press.
Williamson, Oliver, and Winter, Sidney, eds. 1991. *The Nature of the Firm.* Oxford: Oxford University Press.
Wimsatt, William. 1980. "Reductionist Research Strategies and Their Biases in the Units of Selection Controversy." In *Scientific Discovery,* pp. 213–259. See Nickels (1980).
Winch, Peter. 1985. *The Idea of A Social Science.* London: Routledge.
Winterhalder, B., and Smith, E., eds. 1981. *Hunter-Gatherer Foraging Strategies.* Chicago: University of Chicago Press.
Woodward, James. 1989. "The Causal Mechanical Model of Explanation." In *Scientific Explanation,* pp. 357–384. See Kitcher and Salmon (1989).
Wright, Larry. 1973. "Functions." *Philosophical Review* 82: 139–168.
   1976. *Teleological Explanations.* Berkeley: University of California Press.
Yang, Bong. 1987. "The Supply and Demand for Exports for Industrialized Countries: A Disequilibrium Analysis." *Applied Economics* 19: 1137–1148.

Young, Frank. 1962. "The Function of Male Initiation Ceremonies: A Cross Cultural Test of an Alternative Hypothesis." *American Journal of Sociology* 67: 376–396.
Young, R. 1988. "Is Population Ecology a Useful Paradigm for the Study of Organizations?" *American Journal of Sociology* 94: 124.
Zahar, Elie. 1989. *Einstein's Revolution.* LaSalle, Ill.: Open Court.

INDEX

abstraction: in causal explanation, 95, 97; idealization and, 66; scientific method and, 48, 69–70
Achinstein, P., 55, 113
aggregate data, 76, 88, 239–40
agrarian political behavior: agrarian class systems and, 71–2; causal inference and, 87–8; ceteris paribus problem and, 78–9; evidence for, 76–9; hypotheses about, 72–6; law/generalization distinction and, 92; Paige's theory of, 70–80
agricultural supply and demand, 237–8
a priori arguments: see conceptual arguments

background knowledge: explanatory depth, 219–20; interpretation, 206; nonexperimental evidence, 85, 86, 88, 89; in testing, 25–6, 33–5, 52–3, 229; the role of mechanisms and, 182
Barten, A., 235
Bayesianism, 26n5, 54n28
Becker, G., 12, 156–9, 240
belief-desire psychology, 201–4, 226, 230, 238–9
bias, 41, 46, 249, 263
Bigelow, J., 114
Bloor, D., 38
Borse, C., 112–3
Bowles, S., 104
Boyd, R., 140
bridge laws, 147–8
Bronsard, C., 236
Brown, G., 13, 212–5

Carnap, R., 18, 19
Cartwright, N., 64n5, 66n8
case study evidence, 88–90
causal explanation, 10, 11, 55, 81, 83, 94, 96–8, 243; ceteris paribus clauses and, 64–70; free will and, 150; functional explanation and, 106–14, 136; individualism and, 168–70; macrosociological, 172–7; pragmatic factors and, 55–6, 169–70, 178
causal generalizations: versus laws, 90–1; see also explanation
causal inference, 85–8
cell biology: ceteris paribus claims and, 64; as counterexample to unificationist approaches, 166; cross testing and, 54; generic predictions and, 243; as lacking formal theories, 98; theory laden data and, 34
ceteris paribus clauses, 10, 63–80; in economics, 227, 235, 237–8; evolutionary biology and, 82–3; in functional explanations, 124; methods for confirming, 67–9; Paige's theory and, 78–9
Cheney, D., 208
Cloward, R., 104, 108
Cohen, G., 109–11, 137n24
conceptual arguments/analysis, 2, 7, 20, 105, 109n2, 259; individualism-holism and, 149–53; interpretation and, 199, 207; against social laws, 59–63
confirmation, 24; aggregate data and, 183–4; Bayesianism and, 26n5; ceteris paribus clauses and, 10, 67–9; functional explanation and, 11, 114–22; generic predic-

279

confirmation (*cont.*)
tions and, 243; increasing predictive success and, 242; interpretation and, 199, 201–4; mechanisms and, 136–7, 179–81; of supply and demand laws, 232–41; *see also* holism of testing, scientific virtues
context sensitivity, 154, 156, 165, 185
Cummins, R., 106

Darwin, C., 1, 84, 90, 139
Davidson, D., 42, 204n8
Davis, K., 104
Day, T., 93n17, 175n30
Deaton, A., 235
de Marchi, N., 228
depression, 212–15
Dickman, J., 110n3
Downs, A., 12, 159–60
Durkheim, E., 3, 6, 7, 11, 101, 103

Earman, J., 91
ecology, 80–1; nonexperimental evidence and, 86
eliminative reduction, 149, 174
Ellen, R., 105, 140
Elster, J., 104, 137, 179–81
empiricism, 51
Endler, J., 117, 180
equilibrium assumptions: functional explanation and, 123, 131; law of supply and demand and, 236; value neutrality and, 45
evidence: notions of, 27, 29; Quine's restricted notion of, 198; role of subject's categories in, 207–8, 214; *see also* confirmation, scientific virtues
evolutionary biology, 81–4; ceteris paribus problem and, 82–3; explanation by argument patterns, 245; functional explanation and, 112, 117, 140–1; Grant's work on 83–4; mechanisms and, 180n25; nonexperimental evidence and, 86; unificationist view of explanation and, 93, 94–5
expectations: in law of supply and demand, 238, 239; in macroeconomics, 235
explanation, 54–6, 169–70; causal, 10–1, 55, 81, 83, 94, 96–8, 231, 243; ceteris paribus clauses and, 64–70; in economics, 231, 241, 245–7; norms and, 218–9; pragmatic factors in, 55–6, 169–70, 178; role of laws in, 58, 90–2; of social science failures, 258–9, 261–5; subject's categories and, 207, 214; symbolism and, 216–7; unification and, 11, 67, 92–7, 178n31, 216; unity of science and, 171–2

fact-value distinction, *see* value neutrality
Faia, M., 106–7
falsificationism, 24–6
Fay, B., 61–2
Feleppa, R., 195
Feynman, R., 23
Fisk, M., 127
Fleischman, M., 265
folk psychology, 201–4
foundationalism, xvi, 20, 42
Franklin, A., 34, 52
Freeman, J., 11, 12, 104, 131–6, 161–3, 180, 183, 193, 262
Friedman, M., 223–4, 227–8
functional explanation, as circular causation, 106–7; as consequence explanation, 109–112; criticisms of, 11, 104–5, 136–41; distinguished from functionalism, 103, 105–6; homeostatic account of, 107–8; Marxist theory of the state and, 126–30, 136–7, 139; mechanisms for, 101–5, 123–4, 125–6; model of, 111–14; natural selection and, 112, 117, obstacles to testing, 118–22; optimality arguments for, 117–9, 122–6; organizational ecology and, 131–5; role account of, 106; types of evidence for, 115–21
functionalism: criticism of, 104–5; distinguished from functional explanation, 103, 105–6

Galison, P., 52
game theory, 160–1
Geertz, C., 208, 215
general equilibrium theory, 224, 226–7, 232, 233, 241, 244–5, 247, 254–5
general predictions, 242–4
Geyskens, E., 235
Giddens, A., 205
Gintis, H., 104
Glymour, C., 85
Goodman, N., 91
Grant, P., 10, 83–4

Habermas, J., 205
Hallpike, C., 104, 105, 139–40
Hands, W., 229
Hannan, M., 11, 12, 104, 131–6, 161–3, 180, 183, 193, 262
Harris, M., 104, 124–6
Harris, T., 13, 212–5
Hausman, D., 67, 226–7, 244n18, 247–9
Hegel, G., 143
Hellman, G., 110n3, 130
Hempel, C., 24
Henderson, D., 202n5

Index 281

Hindu beef aversion, 124–6
Hirsch, A., 228
holism: aggregate evidence and, 183–4; conceptual arguments for 150–1, 152–3; criticisms of 9, 149–52; economics and, 250–7; empirical evidence for, 153–66; macrosociological causation and, 172–3, 181, 183; main theses, xv, 6–7, 12, 144–5, 153, 161, 177–8; motivations for, 143–4
holism of meaning, 31
holism of testing, 20, 23, 33–5, 46, 52, 53, 85; belief-desire psychology and, 201–4
Homans, G., 155–6
Horan, B., 188n14
hunter-gatherer hunting practices, 123–4
hypothetical-deductivism, 23–4

idealizations: abstractions and, 66; see also ceteris paribus clauses, models
incommensurability: of meaning, 31; of standards, 32
indeterminacy of translation, 196–201
individualism: conceptual arguments for and against, 149–53; in economics, 250–7; empirical evidence against, 153–65; main theses, 145, 167, 175, 179, 182, 187–8; motivations for, 143; as ontological claim, 186–9; as reductionist thesis, 145–66; as thesis about complete explanation, 167–73; as thesis about evidence, 182–4; as thesis about heuristics, 184–87; as thesis about mechanisms, 179–87; truth in, 189–90
inference to the best explanation, 93n17
interpretivism, 8, 13, 205–10

Janssen, M., 251n25
Johnson, C., 130

Kitcher, 29, 31, 92–5, 245
Kuhn, T., 8, 9, 10, 16, 29, 30–7, 48–9

Lakatos, I., 53, 224, 228–9
Latour, B., 16, 40–1
laws, 90–1; behavioral vs. causal, 66–7, 97; case study evidence and, 88, 90; generalizations and, 90–1, 97; scientific explanation and, 58, 90–2; social, 59–63; see also explanation, laws of supply and demand
laws of supply & demand, 232; aggregate nature of, 239–40; as argument pattern, 245–7, ceteris paribus clauses and, 234–6, 237–8; evidence for, 233–8; individualist mechanisms and, 256–7; interpretation of, 233–4; neo-classical theory and, 235n9

Leamer, E., 68
Levins, R., 133
Levi-Strauss, C., 215
Little, D., 104, 179
Longino, H., 46–7, 263

macroeconomics, 252–5
macrosociological causation, 172–3, 181, 183
Marx, K. 1, 2, 3, 6, 11, 75, 101, 103, 126–31, 143, 151, 262
McCloskey, D., 225–6, 229–30
mechanisms: cell biology and, 95; confirmation and, 179–81, 240; economics and, 240, 242, 255–7; evolutionary biology and, 84; explanation and, 96, 181–2; functional explanation and, 104–5, 123–4, 125–6, 136–7, 139; interpretation and, 207; Paige's theory and, 79
Merton, R., 103
methodological individualism, see individualism
microeconomics, 14, 118, 250–1, 254–5; see also laws of supply and demand
microfoundations, 14, 255–7
Miliband, R., 104
Miller, R., 170n25
Mirowski, P., 230n7
models, 99, 262; economics, 231, 247, 249
Montesquieu, B. 3
Moore, S., 104
multiple realizations: as argument against social laws, 59–60; as obstacle to reduction, 152, 155–6, 159, 161–4, 251–3; and reductionist heuristics, 185
Myrdal, G., 45–6

Nagel, E., 107
natural selection, see evolutionary biology
naturalism, 1, 3–4, 27–30, 48, 102, 122, 141, 191–2, 221; ceteris paribus problem and, 70–9, 80–4; consequences of, 5; criticisms of, 8–9, 13, 16; economics and, 222, 238; as explanatory thesis, 56–7; law/generalization distinction and, 191; non-experimental evidence and, 84, 86, 88, 90; Rorty on 42–3; scientific realism and, 29
naturalized epistemology and philosophy of science, 21, 39
Nelson, A., 254
Nelson, R., 251
neo-classical economic theory, 14, 158, 247–50; postulates of, 223; role in supply & demand explanation, 235n9, 240,

neo-classical economic theory (*cont.*) 241; *see also* general equilibrium, microeconomics
Nerlove, M., 238
non-experimental evidence, 10, 84–90; *see also* confirmation
norms, 13, 217–20

optimality arguments, 11, 117–9, 122–6, 137–8, 139
organizational ecology, 131–6

Paige, J., 10, 12, 70–80, 85, 86–8, 89–90, 92, 163–5, 181, 183, 262
Pargetter, R., 114
Parsons, T., 11, 101, 103, 109, 137, 262
physicalism, 151, 197
Piven, F., 104, 108
Pollak, R., 235, 240
Pons, S., 265
Popper, K., 9, 24–6, 143
positivism, 8–9, 17–9, 19–21, 22, 225, 230
post-modernism, 17, 42, 225
prerequisite analysis, 137–8
pseudoscience, 8, 9, 11, 24, 47
Putnam, H., 190n37

Quine, W., 10, 13, 16, 17–21, 191, 194–201

Radcliffe-Brown, A., 3, 11, 103, 106, 109
randomization, 86
Rappaport, R., 104
rational choice theory, 9, 12, 155–61, 219–20
realism: of assumptions, 223–32; about meanings and psychological states, 210–1; naturalism and, 29; *see also* scientific rationality
reduction, 12, 142; conceptual arguments about, 149–53; by elimination, 149, 174–5; empirical evidence against, 153–65; potential obstacles to, 153–5; requirements for, 145–9; supervenience and, 152; unity of science and, 166; 171–2; *see also* individualism
relativism, 8, 9–10
Richardson, R., 147n6
Richerson, P., 140
Rorty, R., 10, 17, 42–3
Rosenberg, A., 152n13, 191, 194, 201–4, 226, 230–1, 238n13, 243–4
Roth, P., 195
Rouse, J., 98n22
Ruben, D., 152–3
Rudner, R., 45–6

Salvas-Bronsard, L., 236
Salmon, M., 64n5
Salmom, W., 182n33
Schrader, D., 251n25
scientific method, 21, 22–6, 27, 32, 41, 42–3; and ceteris paribus problem, 68–90; monistic versus pluralistic conceptions, 57; *see also* confirmation, scientific virtues
scientific rationality, 27–8, 38–41, 48
scientific virtues, 10, 21, 47–56; abstract versus realized, 48, 57, 69–70, 97–8; cross tests and, 53, 69–70, 99, 135, 214, 237, 262; evidential, 49–50, 98–100, 195, 246; fair tests and, 53, 69–70, 85, 89, 99, 135, 206, 214, 262; independent tests and, 52, 69–70, 99, 135, 214, 237, 262; interpretation and, 208, 211, 214; Kuhn on, 35–6; *see also* explanation, theories
Sellers, W., 42
Sensat, J., 147n5
Seyfarth, R., 208
Skocpol, T., 129
Skyrms, B., 91
Smith, A., 3, 243, 244
Smith, E., 104, 123–4
Sober, E., 140
social constructivism, 8, 10, 29, 37–41, 230
social processes in science, 37–41
Soles, D., 195
statistical inference, 46, 261
Stinchcombe, A., 107
Stone, J., 234–5
supervenience, 60, 151–2, 188, 197, 220; explanation and, 168–70
supply and demand argument strategy, 14, 245–7
symbolism, 13, 215–7

Taylor, C., 61–2, 191, 205–6, 209
tendencies, 65–7, 80
Theil, H., 235
theories, 49–50; good science and, 98–100, 246–7, 249; scientific practice and, 36–7
theories of the state, 126–31; individualism and, 164
theory of the test, 25, 33–5, 46, 52
theory-laden data, 33–5
tokens and types, 167–8
Turner, V., 215

unity of science, 142, 166, 171–2, 190; requirements for reduction and, 146–9

value neutrality, 17, 43–7; economics and, 46n21; functional explanation and, 138; social science failure and, 263
van Fraassen, B., 29, 55
Vayda, A., 104, 125

Wales, T., 235, 240

Watkins, J., 151–2
Weber, M., 3
Weintraub, R., 224–5, 228–9
Williamson, O., 104
Winter, S., 251

Young, R., 105

www.ingramcontent.com/pod-product-compliance
Ingram Content Group UK Ltd.
Pitfield, Milton Keynes, MK11 3LW, UK
UKHW032325190125
453752UK00011B/147